Mathematics

Essential Skills

6

253 Normanby Road, South Melbourne, Victoria 3205, Australia

Oxford University Press is a department of the University of Oxford.
It furthers the University's objective of excellence in research, scholarship, and education by publishing worldwide in

Oxford New York

Auckland Cape Town Dar es Salaam Hong Kong Karachi
Kuala Lumpur Madrid Melbourne Mexico City Nairobi
New Delhi Shanghai Taipei Toronto

With offices in

Argentina Austria Brazil Chile Czech Republic France Greece Guatemala Hungary Italy Japan Poland Portugal Singapore South Korea Switzerland Thailand Turkey Ukraine Vietnam

OXFORD is a trade mark of Oxford University Press in the UK and in certain other countries

First published 2010
Reprinted 2014, 2018(D)

ISBN 978 0 19 557403 6

Typeset by Palmer Higgs
Printed and bound in Australia by Ligare Book Printers Pty Ltd

PAPUA NEW GUINEA

Mathematics

Essential Skills

6

Pat Lilburn

OXFORD
UNIVERSITY PRESS
AUSTRALIA & NEW ZEALAND

Contents

For Students

The *Oxford Essential Skills Book* has been written to support the *Oxford Grade 6 Mathematics Student Books A* and *B*.

The focus of this book is on providing practice examples to revise and support the concepts learned in the two Student Books.

The book is set out under the five Strands outlined in the *Upper Primary Mathematics Syllabus* and *Teacher Guide*:

- Number and Application
- Space and Shape
- Measurement
- Chance and Data
- Patterns and Algebra

Within each Strand, you will find many practice examples that relate directly to the Learning Outcomes for that Strand. These Learning Outcomes are specified in the *Upper Primary Mathematics Syllabus.*

The subheading in each Strand shows you what the examples will help you practise.

6.1.1 Add, subtract, multiply and divide fractions

The Help Box contains worked examples to show you how to set out and complete the examples in each Strand. The Remember Box contains information that will help you complete the examples.

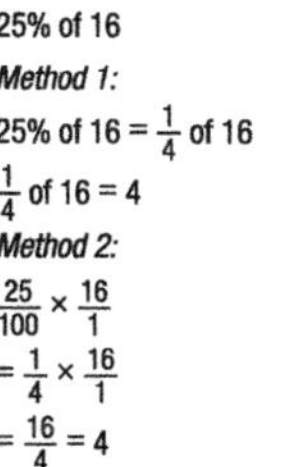

25% of 16

Method 1:

25% of 16 = $\frac{1}{4}$ of 16

$\frac{1}{4}$ of 16 = 4

Method 2:

$\frac{25}{100} \times \frac{16}{1}$

$= \frac{1}{4} \times \frac{16}{1}$

$= \frac{16}{4} = 4$

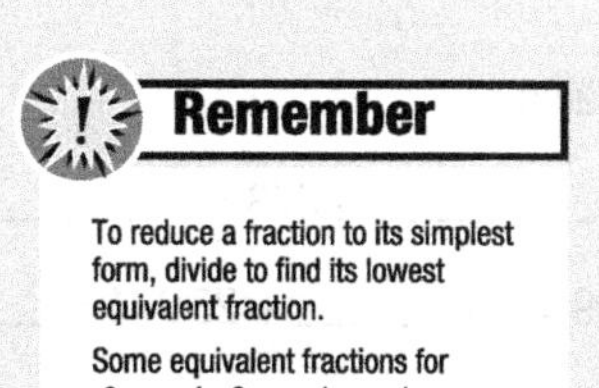

To reduce a fraction to its simplest form, divide to find its lowest equivalent fraction.

Some equivalent fractions for $\frac{8}{16}$ are $\frac{4}{8}$, $\frac{2}{4}$ and $\frac{1}{2}$ but $\frac{1}{2}$ is its lowest equivalent fraction because it cannot be divided any further.

At the end of each Strand there is an Assessment section that covers all the examples from that Strand. If you have difficulty with any test questions, go back to the relevant page in the Strand and look at the worked examples before doing the test again.

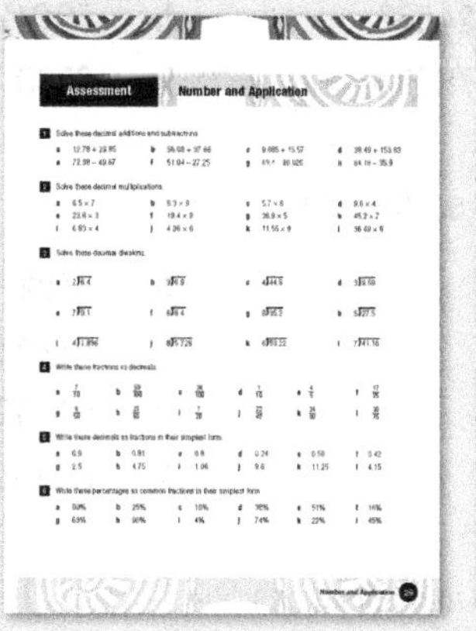

Assessment Number and Application

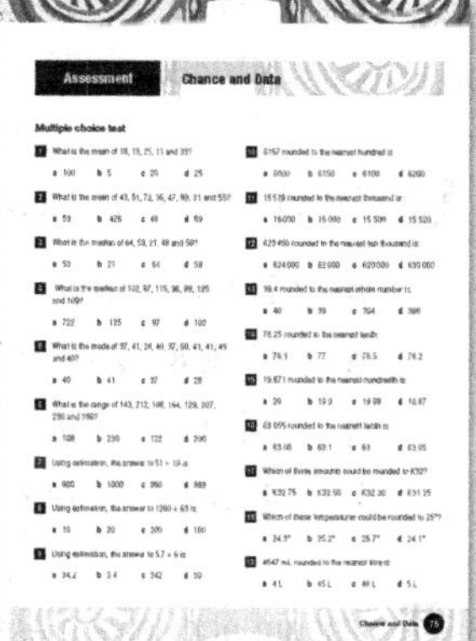

Assessment Chance and Data

Multiple choice test

There is an Answers section and a section for Important Facts at the back of the book. The Answers section contains answers to all practice examples and test questions while the Important Facts section contains some of the facts that you will need to refer to throughout the year.

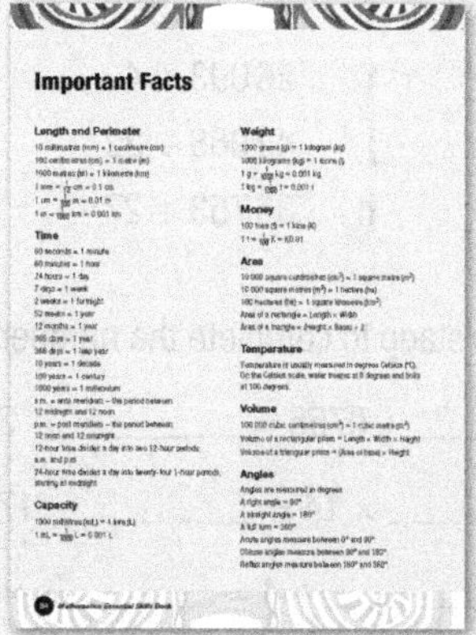

Important Facts

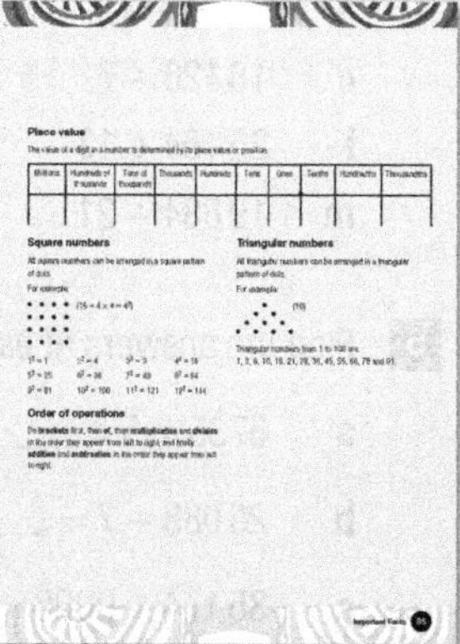

Strand Number and Application

Revise the four operations with whole numbers

1 Set out and solve these additions.

a	6721 + 9648	**b**	3054 + 8979	**c**	5458 + 7207 + 8544
d	4607 + 3886 + 5208	**e**	18 435 + 15 906	**f**	23 067 + 33 755
g	7638 + 14 395 + 24 667	**h**	56 012 + 3694 + 21 826	**i**	9862 + 45 317 + 19 133
j	50 043 + 6966 + 78 299	**k**	6780 + 59 984 + 7785	**l**	33 997 + 5729 + 8903

2 Set out and solve these subtractions.

a	7886 – 5155	**b**	6492 – 3738	**c**	8015 – 5673
d	47 864 – 23 879	**e**	67 818 – 45 946	**f**	80 781 – 56 393
g	74 352 – 7858	**h**	51 112 – 9667	**i**	62 050 – 18 267
j	40 819 – 27 712	**k**	92 245 – 39 884	**l**	78 107 – 49 526

3 Set out and solve these multiplications.

a	5682 × 7	**b**	3365 × 9	**c**	6792 × 6	**d**	4816 × 4
e	27 364 × 8	**f**	12 325 × 5	**g**	32 408 × 3	**h**	56 844 × 7
i	17 663 × 15	**j**	32 281 × 23	**k**	29 894 × 18	**l**	56 098 × 32
m	23 447 × 25	**n**	19 895 × 36	**o**	38 219 × 29	**p**	47 631 × 17

4 Set out and solve these divisions.

a	3729 ÷ 3	**b**	7568 ÷ 2	**c**	5064 ÷ 8	**d**	4955 ÷ 5
e	18 126 ÷ 7	**f**	26 093 ÷ 4	**g**	22 771 ÷ 6	**h**	31 972 ÷ 8
i	25 164 ÷ 13	**j**	42 968 ÷ 22	**k**	32 097 ÷ 15	**l**	27 116 ÷ 31
m	19 884 ÷ 21	**n**	36 753 ÷ 27	**o**	22 908 ÷ 35	**p**	47 154 ÷ 16

5 Find the answers at each stage to complete the number chains.

a 6735 × 5 = ________ + 9756 = ________ ÷ 3 = ________ – 8739 = ________

b 25 088 ÷ 7 = ________ × 14 = ________ – 17 699 = ________ + 23 074 = ________

c 35 174 + 9086 = ________ ÷ 4 = ________ × 9 = ________ – 37 968 = ________

d 88 173 – 54 527 = ________ + 11 987 = ________ ÷ 21 = ________ × 25 = ________

e 5847 × 12 = ________ − 38 129 = ________ ÷ 5 = ________ + 59 942 = ________

f 67 590 ÷ 15 = ________ + 72 389 = ________ − 47 977 = ________ × 7 = ________

g 90 065 − 51 773 = ________ ÷ 3 = ________ × 6 = ________ + 48 947 = ________

h 38 436 + 7519 = ________ − 8947 = ________ ÷ 12 = ________ × 28 = ________

6 Find answers to these problems.

- **a** the product of 3547 and 9
- **b** the total of 87 633 and 32 198
- **c** the difference between 20 058 and 7427
- **d** the quotient of 46 144 and 7
- **e** the sum of 5674 and 3891 and 6602
- **f** the product of 11 377 and 4
- **g** the difference between 8661 and 13 429
- **h** the total of 54 639 and 18 965
- **i** the quotient of 15 and 48 435
- **j** the product of 23 and 3718
- **k** the difference between 6720 and 31 244
- **l** the sum of 65 979 and 8692

Remember

The answer to:

- an addition can be called the **sum** or **total**
- a subtraction can be called the **difference**
- a multiplication can be called the **product**
- a division can be called the **quotient**.

7 Use subtraction to find the missing numbers.

a * * * * * + 26 759 51 095	**b** * * * * * + 17 086 64 425	**c** * * * * * + 34 816 93 958	**d** * * * * * + 19 469 78 275
e 38 135 + * * * * * 67 458	**f** 26 557 + * * * * * 80 727	**g** 43 944 + * * * * * 71 526	**h** 52 273 + * * * * * 93 168

8 Use addition to find the missing numbers.

a * * * * * − 17 592 70 164	**b** * * * * * − 25 481 38 946	**c** * * * * * − 8 759 69 512	**d** * * * * * − 33 508 56 792
e * * * * * − 19 817 48 073	**f** * * * * * − 42 066 38 295	**g** * * * * * − 27 834 54 198	**h** * * * * * − 51 829 26 944

9 Use division to find the missing numbers.

a
$$\begin{array}{r} * \; * * * \\ \times \; 7 \\ \hline 26\;964 \\ \hline \end{array}$$

b
$$\begin{array}{r} * \; * * * \\ \times \; 9 \\ \hline 19\;377 \\ \hline \end{array}$$

c
$$\begin{array}{r} * \; * * * \\ \times \; 6 \\ \hline 35\;682 \\ \hline \end{array}$$

d
$$\begin{array}{r} * \; * * * \\ \times \; 8 \\ \hline 23\;848 \\ \hline \end{array}$$

10 Use multiplication to find the missing numbers.

a
$$\begin{array}{r} 4816 \\ 5\overline{)*\,*\,*\,*\,*} \end{array}$$

b
$$\begin{array}{r} 8527 \\ 4\overline{)*\,*\,*\,*\,*} \end{array}$$

c
$$\begin{array}{r} 6923 \\ 7\overline{)*\,*\,*\,*\,*} \end{array}$$

d
$$\begin{array}{r} 3578 \\ 9\overline{)*\,*\,*\,*\,*} \end{array}$$

Help Box

Equivalent fractions can be made by dividing the numerator and the denominator by the same number.

$$\frac{5 \div 5}{10 \div 5} = \frac{1}{2}$$

So $\frac{5}{10}$ is the same as $\frac{1}{2}$.

Equivalent fractions can also be made by multiplying the numerator and the denominator by the same number.

$$\frac{1 \times 3}{3 \times 3} = \frac{3}{9}$$

So $\frac{1}{3}$ is the same as $\frac{3}{9}$.

6.1.1 Add, subtract, multiply and divide fractions

Make equivalent fractions

1 Use division to write an equivalent fraction for each of these fractions.

a $\frac{2}{4}$ b $\frac{4}{8}$ c $\frac{2}{6}$ d $\frac{2}{10}$

e $\frac{6}{10}$ f $\frac{3}{12}$ g $\frac{5}{15}$ h $\frac{4}{10}$

i $\frac{6}{9}$ j $\frac{5}{20}$ k $\frac{25}{100}$ l $\frac{10}{100}$

m $\frac{3}{18}$ n $\frac{4}{16}$ o $\frac{15}{21}$ p $\frac{3}{9}$

q $\frac{4}{24}$ r $\frac{6}{14}$

2 Use multiplication to write an equivalent fraction for each of these fractions.

a $\frac{1}{4}$ b $\frac{2}{3}$ c $\frac{1}{6}$ d $\frac{4}{5}$ e $\frac{7}{10}$ f $\frac{1}{8}$

g $\frac{3}{7}$ h $\frac{1}{5}$ i $\frac{3}{4}$ j $\frac{5}{8}$ k $\frac{7}{9}$ l $\frac{2}{5}$

m $\frac{5}{7}$ n $\frac{3}{11}$ o $\frac{4}{9}$ p $\frac{5}{12}$ q $\frac{8}{9}$ r $\frac{5}{6}$

3 Fill in the spaces to make an equivalent fraction for each of these fractions.

a $\frac{3}{4} = \frac{}{40}$ b $\frac{5}{12} = \frac{10}{}$ c $\frac{5}{8} = \frac{}{24}$ d $\frac{14}{21} = \frac{2}{}$ e $\frac{6}{15} = \frac{}{5}$

f $\frac{4}{9} = \frac{8}{}$ g $\frac{10}{25} = \frac{2}{}$ h $\frac{2}{3} = \frac{}{12}$ i $\frac{12}{15} = \frac{}{5}$ j $\frac{7}{8} = \frac{42}{}$

4 Write these fractions in their simplest form.

a $\frac{4}{8}$ b $\frac{10}{18}$ c $\frac{8}{24}$ d $\frac{25}{100}$

e $\frac{12}{16}$ f $\frac{9}{15}$ g $\frac{6}{30}$ h $\frac{4}{14}$

i $\frac{8}{10}$ j $\frac{20}{50}$ k $\frac{21}{28}$ l $\frac{9}{27}$

m $\frac{10}{25}$ n $\frac{12}{20}$ o $\frac{18}{24}$ p $\frac{6}{15}$

q $\frac{3}{21}$ r $\frac{5}{60}$

Remember

To reduce a fraction to its simplest form, divide to find its lowest equivalent fraction.

Some equivalent fractions for $\frac{8}{16}$ are $\frac{4}{8}$, $\frac{2}{4}$ and $\frac{1}{2}$.

But $\frac{1}{2}$ is its lowest equivalent fraction because it cannot be divided any further.

5 Write the fractions in each row that are equivalent to the fraction on the left.

a	$\frac{1}{3}$	$\frac{2}{9}$	$\frac{4}{12}$	$\frac{2}{6}$	$\frac{5}{10}$	$\frac{10}{30}$
b	$\frac{5}{6}$	$\frac{25}{30}$	$\frac{1}{3}$	$\frac{10}{15}$	$\frac{7}{8}$	$\frac{15}{18}$
c	$\frac{9}{12}$	$\frac{6}{9}$	$\frac{3}{4}$	$\frac{4}{5}$	$\frac{75}{100}$	$\frac{6}{8}$
d	$\frac{10}{15}$	$\frac{5}{7}$	$\frac{4}{6}$	$\frac{2}{3}$	$\frac{8}{12}$	$\frac{2}{5}$
e	$\frac{3}{8}$	$\frac{6}{10}$	$\frac{8}{16}$	$\frac{12}{32}$	$\frac{9}{24}$	$\frac{30}{80}$
f	$\frac{1}{5}$	$\frac{2}{10}$	$\frac{3}{5}$	$\frac{4}{20}$	$\frac{3}{12}$	$\frac{3}{15}$
g	$\frac{12}{15}$	$\frac{6}{7}$	$\frac{8}{10}$	$\frac{9}{10}$	$\frac{4}{5}$	$\frac{16}{24}$
h	$\frac{2}{7}$	$\frac{10}{35}$	$\frac{6}{21}$	$\frac{4}{9}$	$\frac{8}{25}$	$\frac{14}{49}$
i	$\frac{2}{12}$	$\frac{3}{15}$	$\frac{4}{20}$	$\frac{1}{6}$	$\frac{3}{18}$	$\frac{1}{4}$
j	$\frac{9}{10}$	$\frac{3}{4}$	$\frac{15}{16}$	$\frac{18}{20}$	$\frac{4}{5}$	$\frac{90}{100}$

Mixed numbers and improper fractions

1 Write the improper fractions in this group of fractions and mixed numbers.

$\frac{3}{7}$ $1\frac{5}{8}$ $\frac{15}{7}$ $\frac{10}{10}$ $\frac{1}{3}$ $\frac{9}{4}$ $\frac{11}{7}$ $2\frac{1}{4}$

$\frac{5}{6}$ $4\frac{1}{2}$ $\frac{7}{2}$ $\frac{8}{3}$ $\frac{6}{6}$ $\frac{2}{5}$ $\frac{3}{4}$ $\frac{5}{3}$

$\frac{10}{5}$ $3\frac{2}{3}$ $\frac{8}{6}$ $\frac{7}{12}$ $\frac{19}{4}$ $\frac{7}{8}$

Help Box

An improper fraction is a fraction with a numerator that is greater than or equal to the denominator.
For example, $\frac{9}{5}$, $\frac{13}{4}$ and $\frac{5}{5}$ are improper fractions.

A mixed number is a number that has a whole number and a fraction.
For example, $2\frac{1}{3}$, $5\frac{1}{2}$ and $4\frac{3}{5}$ are mixed numbers.

Remember

To convert an improper fraction to a mixed number, divide the numerator by the denominator.

$\frac{13}{8} = 13 \div 8 = 1\frac{5}{8}$

2 Convert these improper fractions to mixed numbers.

a	$\frac{15}{7}$	**b**	$\frac{7}{2}$	**c**	$\frac{10}{3}$	**d**	$\frac{19}{4}$
e	$\frac{20}{7}$	**f**	$\frac{12}{9}$	**g**	$\frac{13}{5}$	**h**	$\frac{23}{9}$
i	$\frac{5}{3}$	**j**	$\frac{8}{5}$	**k**	$\frac{11}{4}$	**l**	$\frac{14}{3}$
m	$\frac{16}{7}$	**n**	$\frac{21}{6}$	**o**	$\frac{17}{2}$	**p**	$\frac{13}{7}$
q	$\frac{9}{4}$	**r**	$\frac{10}{6}$				

Remember

To convert a mixed number to an improper fraction, multiply the denominator by the whole number. Add the answer to the numerator of the fraction and place the total over the denominator.

For $2\frac{5}{8}$, multiply 8 by 2 (16), add 16 to 5 (21), place 21 over 8: $\frac{21}{8}$

3 Convert these mixed numbers to improper fractions.

a	$5\frac{1}{2}$	**b**	$7\frac{2}{3}$	**c**	$3\frac{4}{5}$	**d**	$1\frac{7}{8}$
e	$4\frac{3}{4}$	**f**	$2\frac{5}{6}$	**g**	$8\frac{1}{4}$	**h**	$2\frac{9}{10}$
i	$6\frac{1}{9}$	**j**	$5\frac{3}{7}$	**k**	$1\frac{4}{9}$	**l**	$3\frac{5}{8}$
m	$9\frac{3}{4}$	**n**	$12\frac{1}{2}$	**o**	$7\frac{3}{8}$	**p**	$6\frac{3}{5}$
q	$4\frac{4}{7}$	**r**	$8\frac{11}{12}$				

4 Which is larger? Convert one fraction in each pair so that you can compare them to find the larger one.

a	$2\frac{5}{9}$ or $\frac{21}{9}$	**b**	$1\frac{3}{4}$ or $\frac{9}{4}$	**c**	$\frac{15}{6}$ or $2\frac{1}{6}$
d	$\frac{22}{10}$ or $2\frac{4}{5}$	**e**	$\frac{19}{3}$ or $6\frac{2}{3}$	**f**	$4\frac{7}{9}$ or $\frac{41}{9}$
g	$3\frac{2}{5}$ or $\frac{14}{5}$	**h**	$7\frac{3}{4}$ or $\frac{29}{4}$	**i**	$\frac{17}{5}$ or $3\frac{3}{5}$
j	$\frac{26}{7}$ or $3\frac{2}{7}$	**k**	$\frac{13}{8}$ or $1\frac{7}{8}$	**l**	$\frac{34}{10}$ or $3\frac{4}{5}$
m	$5\frac{3}{4}$ or $\frac{25}{4}$	**n**	$3\frac{1}{3}$ or $\frac{8}{3}$	**o**	$4\frac{5}{7}$ or $\frac{31}{7}$
p	$\frac{28}{3}$ or $9\frac{2}{3}$	**q**	$\frac{42}{5}$ or $8\frac{1}{5}$	**r**	$\frac{15}{2}$ or $7\frac{3}{4}$
s	$3\frac{5}{6}$ or $\frac{25}{6}$	**t**	$4\frac{1}{9}$ or $\frac{35}{9}$	**u**	$3\frac{3}{8}$ or $\frac{29}{8}$

Add and subtract common fractions

Help Box

Add and subtract fractions with like denominators:

- $\frac{2}{7} + \frac{3}{7} = \frac{5}{7}$
- $\frac{9}{10} - \frac{5}{10} = \frac{4}{10}$ or $\frac{2}{5}$

Add and subtract fractions with unlike denominators:

- $\frac{5}{6} - \frac{2}{3} = \frac{5}{6} - \frac{4}{6} = \frac{1}{6}$
- $\frac{2}{3} + \frac{1}{4} = \frac{8}{12} + \frac{3}{12} = \frac{11}{12}$

1 These additions and subtractions have like denominators. Convert any answers that are improper fractions to mixed numbers, and write all answers in their simplest form.

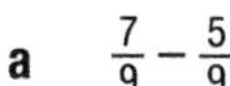

a $\frac{7}{9} - \frac{5}{9}$ **b** $\frac{2}{3} + \frac{2}{3}$ **c** $\frac{2}{7} + \frac{4}{7}$

d $\frac{5}{8} - \frac{3}{8}$ **e** $\frac{5}{6} + \frac{5}{6}$ **f** $\frac{9}{11} - \frac{4}{11}$

g $\frac{17}{20} - \frac{8}{20}$ **h** $\frac{7}{10} + \frac{4}{10}$ **i** $\frac{13}{15} - \frac{7}{15}$

j $\frac{4}{5} + \frac{4}{5} + \frac{3}{5}$ **k** $\frac{1}{4} + \frac{3}{4} + \frac{3}{4}$ **l** $\frac{11}{12} - \frac{6}{12}$

m $\frac{2}{7} + \frac{6}{7} + \frac{5}{7}$ **n** $\frac{7}{8} - \frac{3}{8}$ **o** $\frac{5}{6} - \frac{1}{6}$

p $\frac{5}{8} + \frac{2}{8} + \frac{7}{8}$

2 These additions and subtractions have unlike denominators. Convert any answers that are improper fractions to mixed numbers, and write all answers in their simplest form.

a $\frac{3}{10} + \frac{11}{100}$ **b** $\frac{8}{10} - \frac{2}{5}$ **c** $\frac{5}{8} - \frac{1}{4}$ **d** $\frac{4}{5} + \frac{6}{10}$

e $\frac{5}{6} - \frac{2}{3}$ **f** $\frac{5}{6} + \frac{3}{12}$ **g** $\frac{3}{4} + \frac{7}{12}$ **h** $\frac{11}{15} - \frac{1}{5}$

i $\frac{2}{3} + \frac{1}{2}$ **j** $\frac{3}{4} - \frac{2}{3}$ **k** $\frac{1}{3} - \frac{2}{7}$ **l** $\frac{7}{8} + \frac{2}{3}$

m $\frac{5}{8} - \frac{1}{5}$ **n** $\frac{2}{5} + \frac{3}{4}$ **o** $\frac{3}{8} + \frac{1}{6}$ **p** $\frac{11}{12} - \frac{3}{5}$

3 The answers to these calculations are written below the grid. Write the letter that matches each answer to discover the secret word.

E $\frac{1}{8} + \frac{5}{8}$	**X** $\frac{1}{5} + \frac{1}{4}$	**L** $\frac{7}{12} - \frac{1}{2}$
E $\frac{3}{4} - \frac{2}{5}$	**T** $\frac{1}{3} + \frac{3}{10}$	**E** $\frac{3}{5} - \frac{1}{3}$
N $\frac{2}{5} + \frac{1}{2}$	**L** $\frac{7}{8} - \frac{3}{5}$	**C** $\frac{3}{8} + \frac{5}{12}$

$\frac{4}{15}$ $\frac{9}{20}$ $\frac{19}{24}$ $\frac{7}{20}$ $\frac{1}{12}$ $\frac{11}{40}$ $\frac{3}{4}$ $\frac{9}{10}$ $\frac{19}{30}$

____ ____ ____ ____ ____ ____ ____ ____ ____

Add mixed numbers

Help Box

Add fractions with like denominators:

$2\frac{3}{5} + 3\frac{4}{5} = 5 + (\frac{3}{5} + \frac{4}{5})$

$= 5 + \frac{7}{5} = 5 + 1\frac{2}{5} = 6\frac{2}{5}$

Add fractions with unlike denominators:

$1\frac{3}{4} + 2\frac{5}{12} = 1 + 2 + \frac{3}{4} + \frac{5}{12}$

$= 3 + (\frac{3}{4} + \frac{5}{12}) = 3 + (\frac{9}{12} + \frac{5}{12})$

$= 3 + \frac{14}{12} = 3 + 1\frac{2}{12} = 4\frac{2}{12} = 4\frac{1}{6}$

Add these mixed numbers to see how many fish you can catch.

1 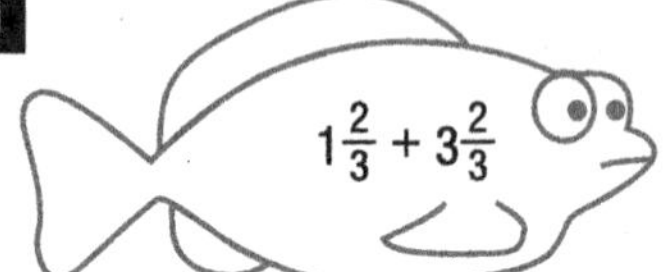 $1\frac{2}{3} + 3\frac{2}{3}$

2 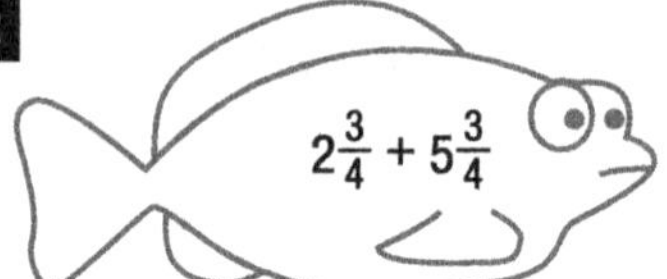 $2\frac{3}{4} + 5\frac{3}{4}$

3 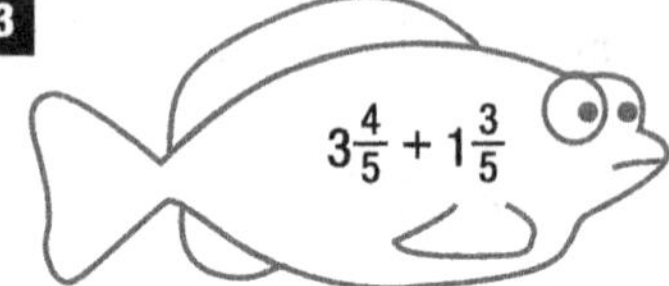 $3\frac{4}{5} + 1\frac{3}{5}$

4 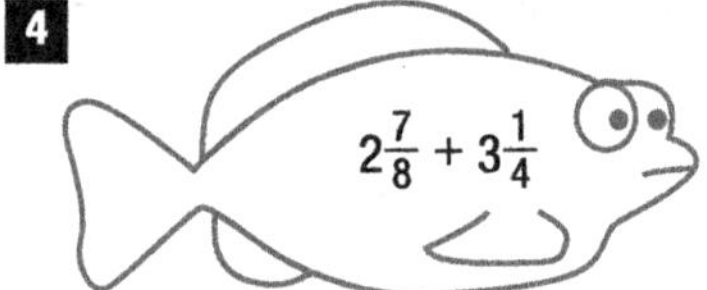 $2\frac{7}{8} + 3\frac{1}{4}$

5 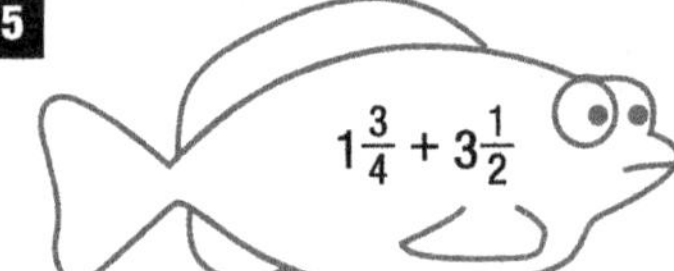 $1\frac{3}{4} + 3\frac{1}{2}$

6 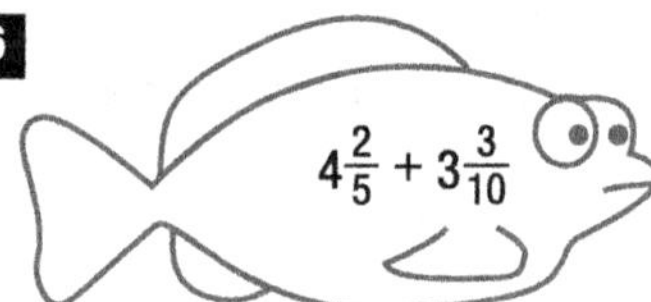 $4\frac{2}{5} + 3\frac{3}{10}$

7 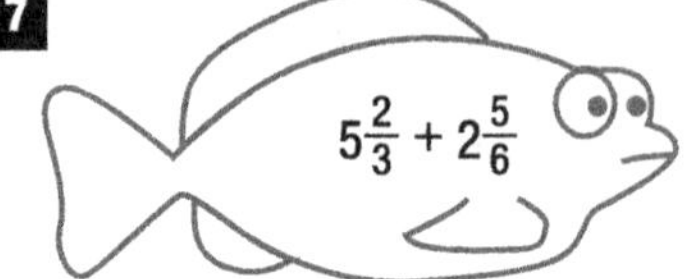 $5\frac{2}{3} + 2\frac{5}{6}$

8 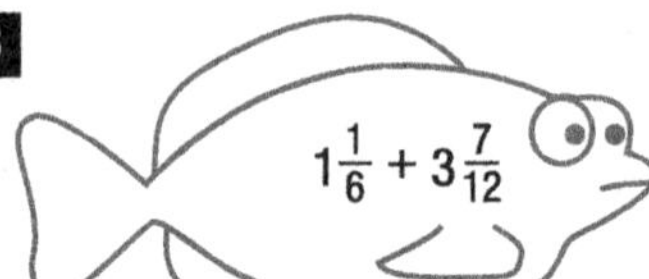 $1\frac{1}{6} + 3\frac{7}{12}$

9 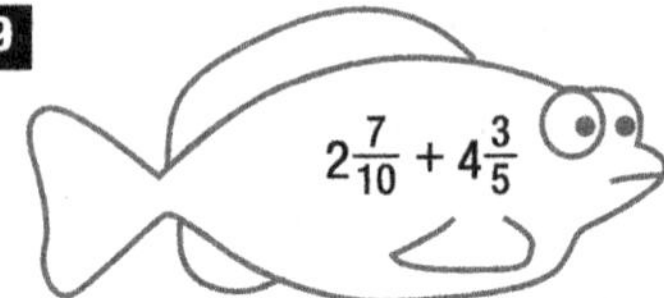 $2\frac{7}{10} + 4\frac{3}{5}$

10 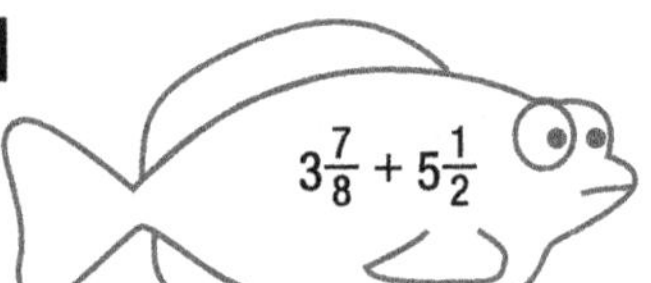 $3\frac{7}{8} + 5\frac{1}{2}$

11 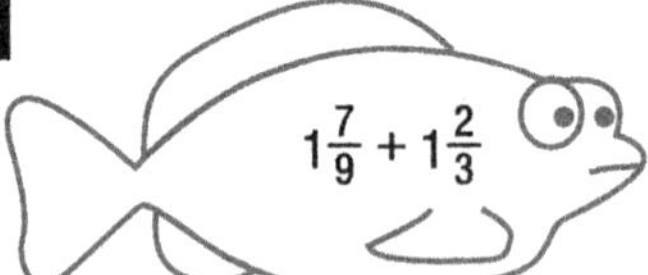 $1\frac{7}{9} + 1\frac{2}{3}$

12 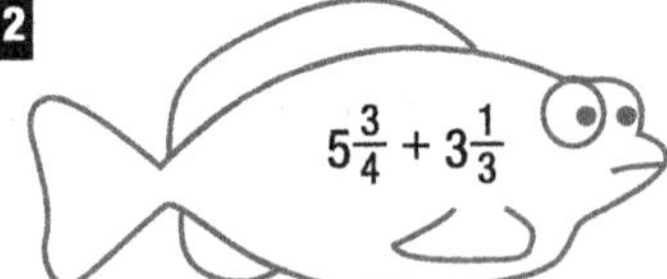 $5\frac{3}{4} + 3\frac{1}{3}$

13 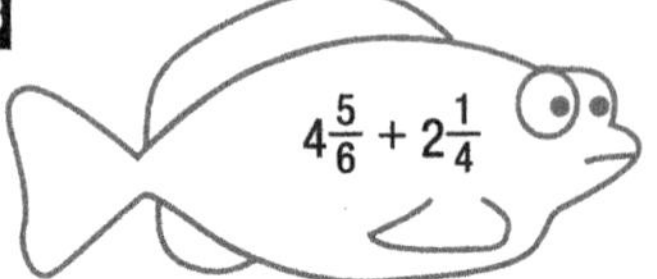 $4\frac{5}{6} + 2\frac{1}{4}$

14 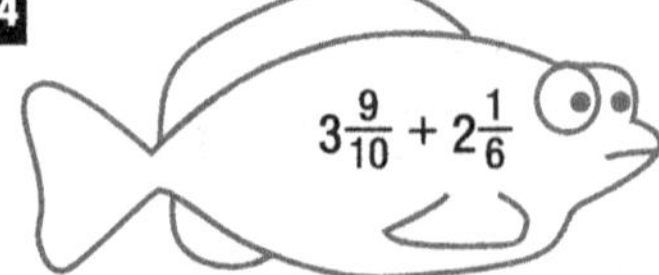 $3\frac{9}{10} + 2\frac{1}{6}$

15 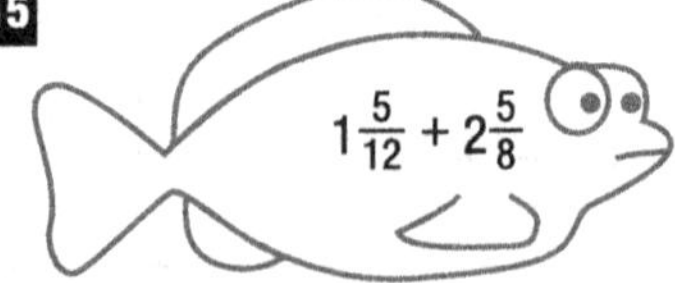 $1\frac{5}{12} + 2\frac{5}{8}$

16 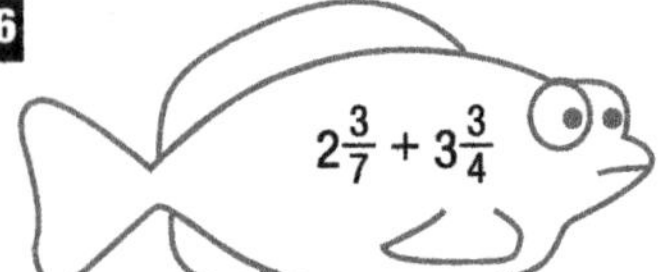 $2\frac{3}{7} + 3\frac{3}{4}$

17 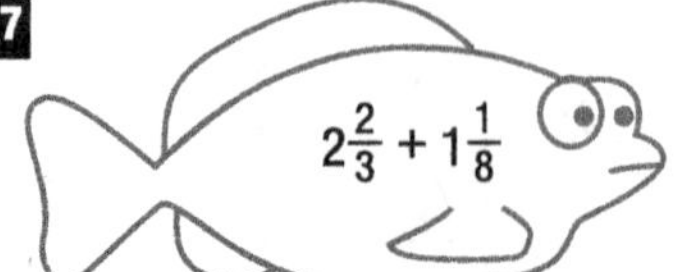 $2\frac{2}{3} + 1\frac{1}{8}$

18 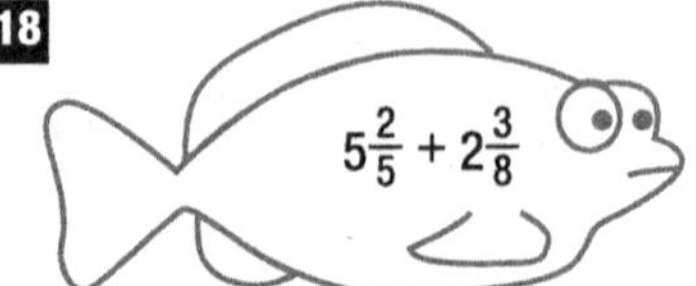 $5\frac{2}{5} + 2\frac{3}{8}$

How many fish did you catch? ________

Subtract mixed numbers

Help Box

Subtract fractions with like and unlike denominators: $4\frac{3}{5} - 2\frac{3}{4} = \frac{23}{5} - \frac{11}{4} = \frac{92}{20} - \frac{55}{20} = \frac{37}{20} = 1\frac{17}{20}$

Subtract these mixed numbers to see how many coconuts you can crack.

1 $3\frac{3}{8} - 1\frac{7}{8}$

2 $6\frac{1}{10} - 3\frac{3}{10}$

3 $5\frac{5}{12} - 2\frac{11}{12}$

4 $4\frac{7}{10} - 1\frac{4}{5}$

5 $3\frac{1}{4} - 1\frac{5}{8}$

6 $5\frac{2}{3} - 2\frac{5}{6}$

7 $2\frac{4}{9} - 1\frac{2}{3}$

8 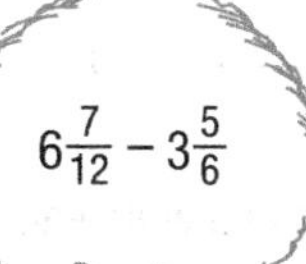 $6\frac{7}{12} - 3\frac{5}{6}$

9 $4\frac{1}{2} - 2\frac{9}{10}$

10 $5\frac{1}{4} - 3\frac{5}{12}$

11 $3\frac{2}{5} - 1\frac{3}{10}$

12 $4\frac{5}{7} - 2\frac{1}{3}$

13 $2\frac{1}{5} - 1\frac{2}{3}$

14 $7\frac{3}{4} - 4\frac{1}{5}$

15 $5\frac{1}{8} - 2\frac{1}{6}$

16 $3\frac{5}{9} - 1\frac{1}{2}$

17 $4\frac{1}{12} - 2\frac{6}{7}$

18 $6\frac{3}{10} - 4\frac{2}{3}$

How many coconuts did you crack? ________

Multiply fractions

Help Box

Multiply by a whole number:

$7 \times \frac{1}{4}$

$= \frac{7}{1} \times \frac{1}{4}$

$= \frac{7}{4}$

$= 1\frac{3}{4}$

Fractions of groups:

$\frac{2}{3}$ of 15

$= \frac{2}{3} \times \frac{15}{1}$

$= \frac{30}{3}$

$= 10$

Multiply a fraction by a fraction:

$\frac{1}{4} \times \frac{2}{5}$

$= \frac{2}{20}$

$= \frac{1}{10}$

Multiply mixed numbers:

$1\frac{1}{4} \times 2\frac{2}{3}$

$= \frac{5}{4} \times \frac{8}{3}$

$= \frac{40}{12}$

$= 3\frac{4}{12}$

$= 3\frac{1}{3}$

1 Multiply these fractions and whole numbers. Write each answer in its simplest form.

a	$5 \times \frac{1}{3}$	**b**	$\frac{1}{5} \times 8$	**c**	$4 \times \frac{3}{4}$	**d**	$\frac{2}{3} \times 3$
e	$2 \times \frac{7}{8}$	**f**	$6 \times \frac{3}{5}$	**g**	$\frac{5}{6} \times 3$	**h**	$7 \times \frac{3}{8}$
i	$5 \times \frac{4}{7}$	**j**	$\frac{2}{5} \times 9$	**k**	$\frac{4}{9} \times 4$	**l**	$5 \times \frac{2}{7}$
m	$\frac{3}{10} \times 2$	**n**	$8 \times \frac{4}{5}$	**o**	$6 \times \frac{5}{8}$		

2 Find the fraction of each number.

a	$\frac{3}{4}$ of 20	**b**	$\frac{2}{5}$ of 15	**c**	$\frac{2}{3}$ of 12	**d**	$\frac{1}{6}$ of 24
e	$\frac{7}{10}$ of 60	**f**	$\frac{1}{3}$ of 24	**g**	$\frac{5}{6}$ of 18	**h**	$\frac{2}{7}$ of 21
i	$\frac{4}{5}$ of 10	**j**	$\frac{4}{9}$ of 27	**k**	$\frac{3}{10}$ of 30	**l**	$\frac{1}{4}$ of 100
m	$\frac{5}{8}$ of 40	**n**	$\frac{3}{7}$ of 35	**o**	$\frac{8}{9}$ of 90		

3 Calculate and write each answer in its simplest form.

a	$\frac{1}{4} \times \frac{2}{5}$	**b**	$\frac{1}{2} \times \frac{2}{3}$	**c**	$\frac{1}{3} \times \frac{4}{5}$	**d**	$\frac{3}{10} \times \frac{1}{6}$
e	$\frac{2}{5} \times \frac{5}{6}$	**f**	$\frac{2}{7} \times \frac{3}{4}$	**g**	$\frac{1}{8} \times \frac{1}{4}$	**h**	$\frac{2}{3} \times \frac{6}{7}$
i	$\frac{3}{8} \times \frac{1}{3}$	**j**	$\frac{5}{7} \times \frac{9}{10}$	**k**	$\frac{3}{4} \times \frac{4}{9}$	**l**	$\frac{1}{2} \times \frac{5}{8}$
m	$\frac{5}{12} \times \frac{2}{5}$	**n**	$\frac{4}{7} \times \frac{3}{10}$	**o**	$\frac{7}{8} \times \frac{3}{5}$		

4 Calculate and write each answer in its simplest form.

a	$2\frac{1}{3} \times \frac{3}{4}$	**b**	$3\frac{1}{2} \times \frac{2}{3}$	**c**	$\frac{1}{2} \times 4\frac{1}{8}$	**d**	$1\frac{1}{4} \times 2\frac{3}{4}$
e	$2\frac{4}{5} \times 1\frac{1}{5}$	**f**	$3\frac{5}{6} \times 2\frac{1}{2}$	**g**	$1\frac{7}{8} \times 1\frac{2}{5}$	**h**	$4\frac{3}{10} \times 2\frac{2}{5}$
i	$2\frac{5}{8} \times 3\frac{1}{6}$	**j**	$1\frac{2}{3} \times 3\frac{3}{5}$	**k**	$1\frac{3}{4} \times 1\frac{2}{7}$	**l**	$3\frac{1}{3} \times 2\frac{1}{2}$
m	$2\frac{5}{9} \times 1$	**n**	$3\frac{1}{7} \times 1\frac{3}{8}$	**o**	$4\frac{1}{2} \times 1\frac{7}{8}$		

5 How many balloons can you burst?

a

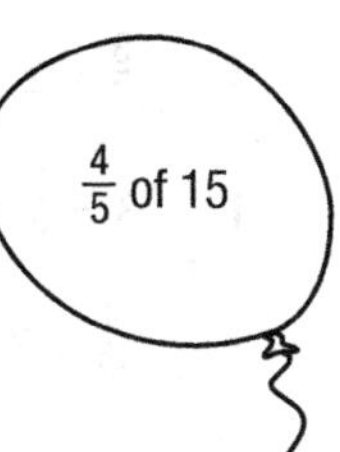

b

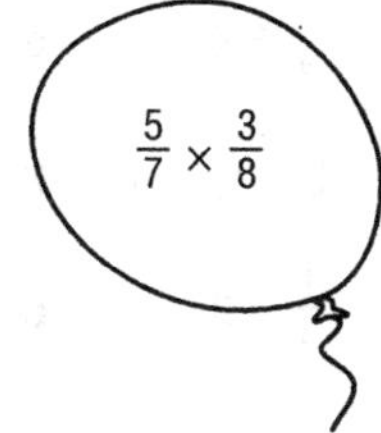

c

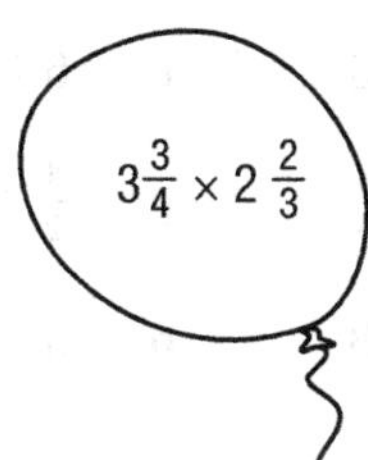

d

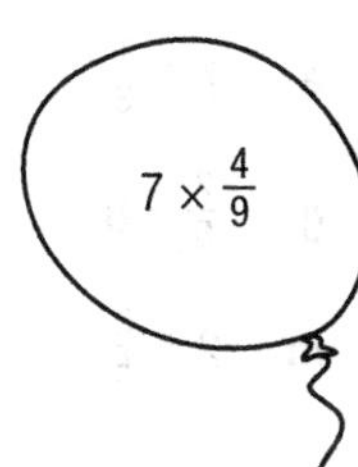

e

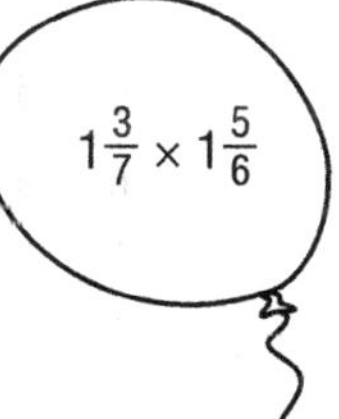

f

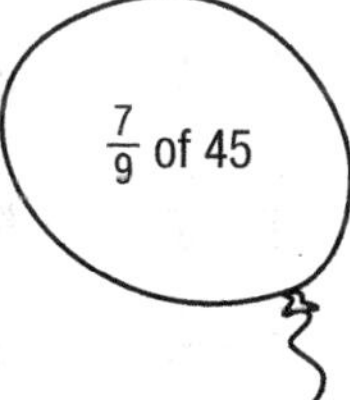

g

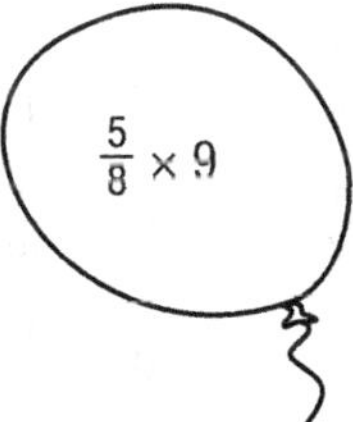

h

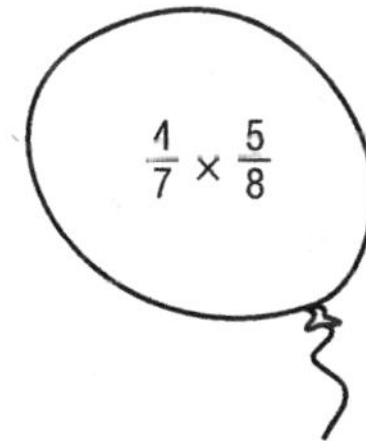

i

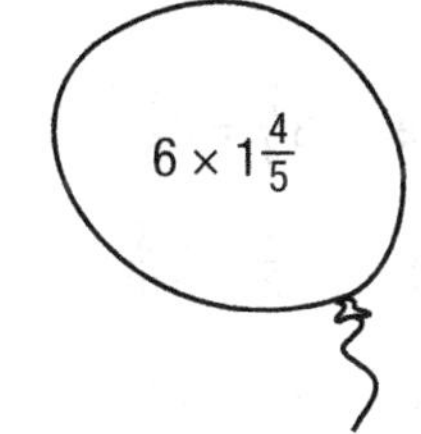

j

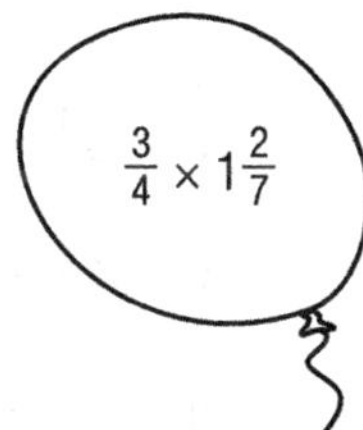

k

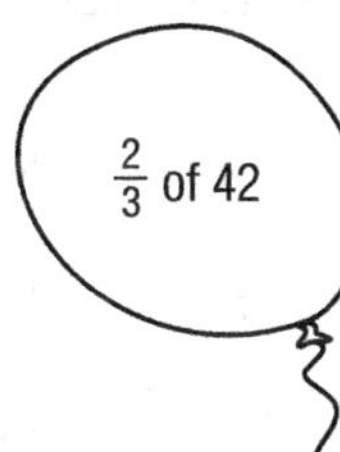

l

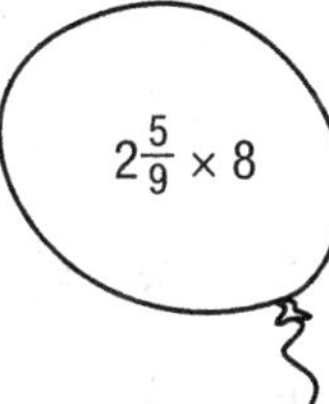

How many balloons did you burst? ______

Divide fractions

Help Box

Divide a whole number by a fraction:

$5 \div \frac{1}{4}$
$= \frac{5}{1} \div \frac{1}{4}$
$= \frac{5}{1} \times \frac{4}{1}$
$= \frac{20}{1}$
$= 20$

Divide a fraction by a whole number:

$\frac{4}{5} \div 3$
$= \frac{4}{5} \div \frac{3}{1}$
$= \frac{4}{5} \times \frac{1}{3}$
$= \frac{4}{15}$

Divide a fraction by a fraction:

$\frac{3}{4} \div \frac{1}{5}$
$= \frac{3}{4} \times \frac{5}{1}$
$= \frac{15}{4}$
$= 3\frac{3}{4}$

Divide mixed numbers:

$2\frac{1}{3} \div 1\frac{1}{2}$
$= \frac{7}{3} \div \frac{3}{2}$
$= \frac{7}{3} \times \frac{2}{3}$
$= \frac{14}{9}$
$= 1\frac{5}{9}$

1 Divide and write each answer in its simplest form.

a $7 \div \frac{1}{3}$	**b** $\frac{5}{6} \div 3$	**c** $6 \div \frac{7}{8}$	**d** $\frac{2}{9} \div 4$	**e** $5 \div \frac{3}{4}$	**f** $\frac{3}{7} \div 2$
g $\frac{2}{3} \div 8$	**h** $4 \div \frac{3}{8}$	**i** $\frac{1}{2} \div 7$	**j** $9 \div \frac{3}{5}$	**k** $\frac{5}{7} \div 6$	**l** $3 \div \frac{5}{9}$
m $2 \div \frac{2}{5}$	**n** $\frac{5}{8} \div 8$	**o** $5 \div \frac{8}{9}$	**p** $\frac{1}{6} \div 3$	**q** $8 \div \frac{7}{10}$	**r** $\frac{5}{12} \div 7$

2 Calculate and write each answer in its simplest form.

a $\frac{2}{3} \div \frac{4}{5}$	**b** $\frac{3}{7} \div \frac{1}{8}$	**c** $\frac{3}{4} \div \frac{5}{8}$	**d** $\frac{1}{7} \div \frac{1}{3}$	**e** $\frac{2}{5} \div \frac{1}{4}$	**f** $\frac{7}{8} \div \frac{2}{9}$
g $\frac{1}{3} \div \frac{2}{7}$	**h** $\frac{5}{6} \div \frac{1}{2}$	**i** $\frac{3}{10} \div \frac{3}{4}$	**j** $\frac{1}{6} \div \frac{2}{3}$	**k** $\frac{4}{5} \div \frac{2}{5}$	**l** $\frac{6}{7} \div \frac{3}{8}$
m $\frac{7}{10} \div \frac{5}{8}$	**n** $\frac{5}{7} \div \frac{1}{3}$	**o** $\frac{4}{9} \div \frac{3}{4}$	**p** $\frac{2}{3} \div \frac{8}{11}$	**q** $\frac{7}{12} \div \frac{1}{4}$	**r** $\frac{1}{2} \div \frac{9}{10}$

3 Calculate and write each answer in its simplest form.

a $1\frac{4}{7} \div 1\frac{1}{4}$	**b** $2\frac{1}{2} \div 1\frac{5}{8}$	**c** $3\frac{3}{4} \div 1\frac{2}{3}$	**d** $5\frac{1}{3} \div 2\frac{3}{5}$	**e** $3\frac{7}{8} \div 2\frac{1}{5}$
f $4\frac{1}{6} \div 2\frac{3}{4}$	**g** $3\frac{5}{8} \div \frac{3}{10}$	**h** $1\frac{2}{7} \div \frac{1}{8}$	**i** $2\frac{8}{9} \div 1\frac{1}{3}$	**j** $1\frac{2}{3} \div 2\frac{3}{10}$
k $2\frac{1}{9} \div \frac{4}{5}$	**l** $3\frac{5}{7} \div 2\frac{7}{8}$	**m** $2\frac{2}{5} \div \frac{5}{6}$	**n** $1\frac{7}{12} \div 1\frac{1}{2}$	**o** $3\frac{1}{4} \div 2\frac{1}{7}$

4 Can you climb to the top of each ladder by correctly solving the divisions?

a

$2\frac{6}{7} \div \frac{1}{5}$

$\frac{5}{9} \div \frac{2}{3}$

$\frac{3}{4} \div 8$

$\frac{5}{6} \div 1\frac{1}{3}$

$7\frac{1}{2} \div 3\frac{1}{4}$

b

$1\frac{9}{11} \div 1\frac{1}{4}$

$4 \div \frac{11}{12}$

$1\frac{3}{8} \div \frac{2}{3}$

$10 \div 2\frac{1}{8}$

$\frac{9}{20} \div 5$

c

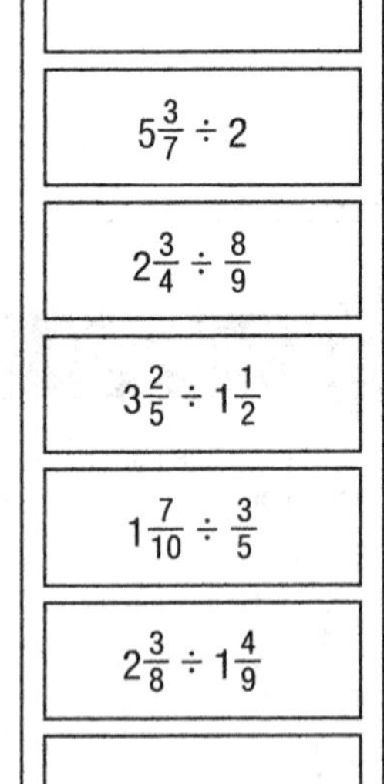

6.1.2. Add, subtract, multiply and divide decimals

Add and subtract decimals

Help Box

Add or subtract decimals with equal numbers of places: 5.47 + 7.85

```
  1 1
  5.47
+ 7.85
 13.32
```

Add or subtract decimals with unequal numbers of places: 25.32 – 8.4

```
 1 14 1
 2̸5̸.32
–  8.40
  16.92
```

Remember

When adding and subtracting decimals, keep the decimal points under one another.

1 Set out and solve these decimal additions.

a	32.5 + 15.4	**b**	5.65 + 2.88	**c**	42.9 + 37.4	**d**	2.17 + 3.46
e	63.82 + 19.59	**f**	25.87 + 18.34	**g**	8.137 + 5.298	**h**	7.649 + 8.073
i	35.6 + 27.14	**j**	16.85 + 21.7	**k**	27.4 + 5.68	**l**	13.85 + 3.672
m	58.725 + 29.32	**n**	124.8 + 78.56	**o**	47.24 + 7.068	**p**	87.9 + 123.85

2 Set out and solve these decimal subtractions.

a	47.8 – 26.5	**b**	9.63 – 4.77	**c**	68.04 – 36.27	**d**	72.16 – 29.38
e	8.758 – 3.964	**f**	25.244 – 15.673	**g**	347.5 – 265.8	**h**	9.152 – 6.709
i	78.22 – 30.7	**j**	81.06 – 35.8	**k**	63.5 – 24.76	**l**	124.3 – 78.42
m	95.2 – 37.658	**n**	145.03 – 86.5	**o**	87.06 – 59.4	**p**	55.802 – 9.14

3 Find answers to these problems:

- **a** the sum of 56.8, 7.925 and 33.19
- **b** the difference between 74.08 and 27.8
- **c** the difference between 80.5 and 116.23
- **d** the total of 137.26, 63.8 and 150.3
- **e** the sum of 345.62 and 67.815
- **f** the difference between 172.4 and 9.06
- **g** the difference between 85.25 and 231.6
- **h** the total of 7.173, 54.8 and 67.06

4 Copy and complete each grid.

a

+	12.6	5.07	3.92	4.552	1.8
4.5					
2.09					
11.8					
5.075					
31.7					

b

–	43.2	17.81	173.5	4.086	87.3
27.5					
9.07					
37.98					
6.154					
118.6					

5 Can you cross the river by correctly solving these calculations?

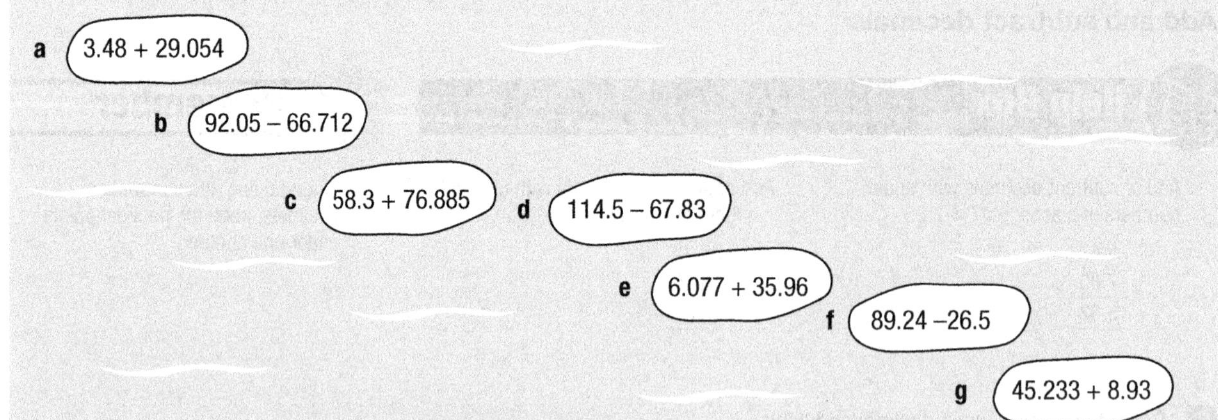

Copy the first set of calculations into your book. Work out the answer to the first calculation. Put this answer in the second calculation and work out its answer. Continue like this until you reach the last calculation. Solve it to check if you are correct. Repeat for each set of calculations.

	6	**7**	**8**	**9**
a	36.172 + 43.86 = __.__	76.003 − 27.64 = __.__	18.94 + 36.725 = __.__	81.32 − 27.068 = __.__
b	[.] − 45.16 = __.__	24.3 + [.] = __.__	[.] − 8.72 = __.__	73.59 − [.] = __.__
c	15.375 + [.] = __.__	[.] − 21.07 = __.__	17.48 + [.] = __.__	[.] + 22.54 = __.__
d	34.89 + [.] = __.__	41.65 + [.] = __.__	31.702 + [.] = __.__	[.] − 7.95 = __.__
e	[.] − 7.63 = **77.507**	[.] − 48.96 = **44.283**	[.] − 9.83 = **86.297**	46.245 + [.] = **80.173**

10 a
  54.03
+ 36.558
______

b
  [.]
− 57.76
______

c
  [.]
+ 9.67
______

d
  47.66
+ [.]
______

e
  [.]
− 57.864
32.294

11 a
  81.322
− 53.09
______

b
  19.58
+ [.]
______

c
  [.]
− 8.09
______

d
  [.]
+ 56.88
______

e
  142.07
− [.]
45.468

12 a
  49.7
+ 78.39
______

b
  [.]
+ 36.5
______

c
  [.]
− 78.612
______

d
  30.08
+ [.]
______

e
  [.]
+ 53.87
169.928

13 a
  95.4
− 17.81
______

b
  113.2
− [.]
______

c
  [.]
+ 128.9
______

d
  [.]
− 89.73
______

e
  33.6
+ [.]
108.38

Multiply decimals

1 Solve these decimal multiplications.

a 4.7 × 5
b 6.3 × 6
c 3.7 × 4
d 5.8 × 6
e 3.5 × 3

f 4.9 × 4
g 12.3 × 7
h 23.9 × 5
i 16.4 × 4
j 33.8 × 2

k 27.6 × 6
l 41.5 × 3
m 58.5 × 8
n 80.7 × 9
o 93.8 × 3

p 64.7 × 6
q 49.3 × 7
r 75.8 × 5

Help Box

Calculate as a normal multiplication and then estimate by rounding to help you place the decimal point in each answer.

Examples:

$^{2}5.7 \times 3 = 17.1$ — Round 5.7 to 6 and multiply by 3 = 18.

$^{3}1^{3}7.^{1}83 \times 4 = 71.32$ — Round 17.83 to 18 and multiply by 4 = 72.

$^{1}3.^{4}1^{1}62 \times 7 = 22.134$ — Round 3.162 to 3 and multiply by 7 = 21.

2 Work out each price tag.

a K3.26 × 4

b K5.17 × 3

c K4.85 × 4

d K6.36 × 5

e K4.73 × 6

f K2.87 × 5

g K5.95 × 7

h K3.66 × 8

i K1.79 × 8

j K6.25 × 3

k K7.52 × 4

l K3.88 × 9

m K7.50 × 7

n K8.67 × 5

o K5.49 × 6

p K9.38 × 4

3 Set out and solve each multiplication.

a	8.9 × 6	**b**	9.3 × 4	**c**	6.96 × 3	**d**	13.7 × 5	**e**	24.8 × 7
f	5.37 × 8	**g**	4.252 × 4	**h**	3.798 × 3	**i**	12.54 × 7	**j**	9.877 × 9
k	5.673 × 5	**l**	35.8 × 6	**m**	8.901 × 7	**n**	15.09 × 5	**o**	7.125 × 6
p	48.7 × 4	**q**	28.16 × 8	**r**	7.09 × 3	**s**	36.18 × 4	**t**	9.67 × 8
u	2.785 × 5	**v**	55.9 × 3	**w**	37.16 × 7	**x**	8.83 × 6		

4 Complete this table.

	Number of items and cost per item	Total cost	Change from K50
a	4 @ K6.70 each		
b	3 @ K11.35 each		
c	5 @ K7.50 each		
d	10 @ K4.25 each		
e	3 @ K9.60 each		

	Number of items and cost per item	Total cost	Change from K50
f	6 @ K7.20 each		
g	4 @ K3.95 each		
h	8 @ K5.35 each		
i	6 @ K6.55 each		
j	5 @ K9.25 each		

Divide decimals

Help Box

Calculate as a normal division and then estimate by rounding to help you place the decimal point in each answer.

Examples: $2\overline{)6.4}$ = 3.2 — Round 6.4 to 6 and divide by 2. The answer is 3 so put the decimal point after 3.

$5\overline{)6.55}$ = 1.31 — Round 6.55 to 7 and divide by 5. The answer is between 1 and 2 so put the decimal point after 1.

Divide these decimal numbers to see how many goals you can shoot.

1 

2 

3 

4 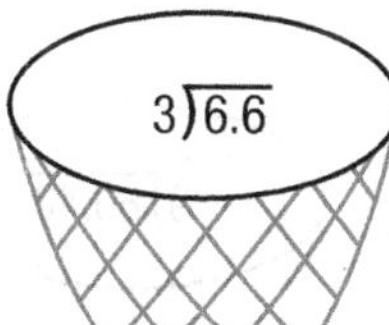

5 

6 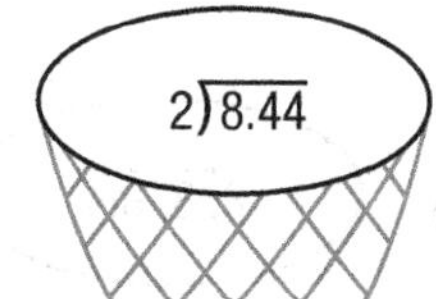

7 

8 

9 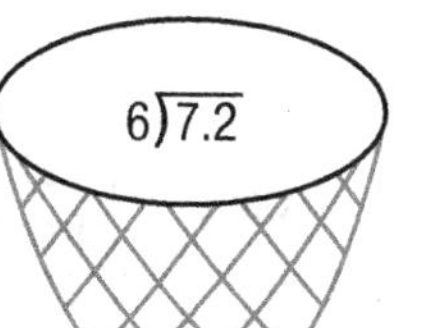

10 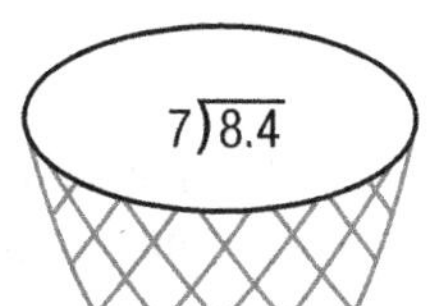

11 

12 

13 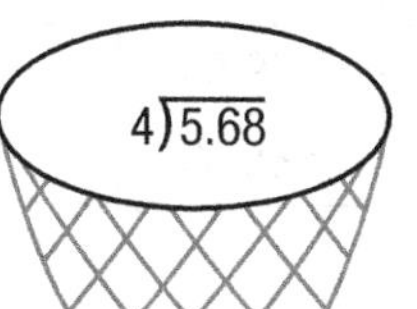

14 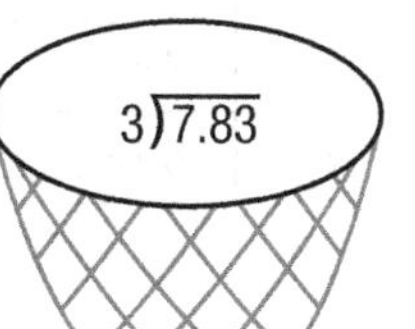

15 

16 

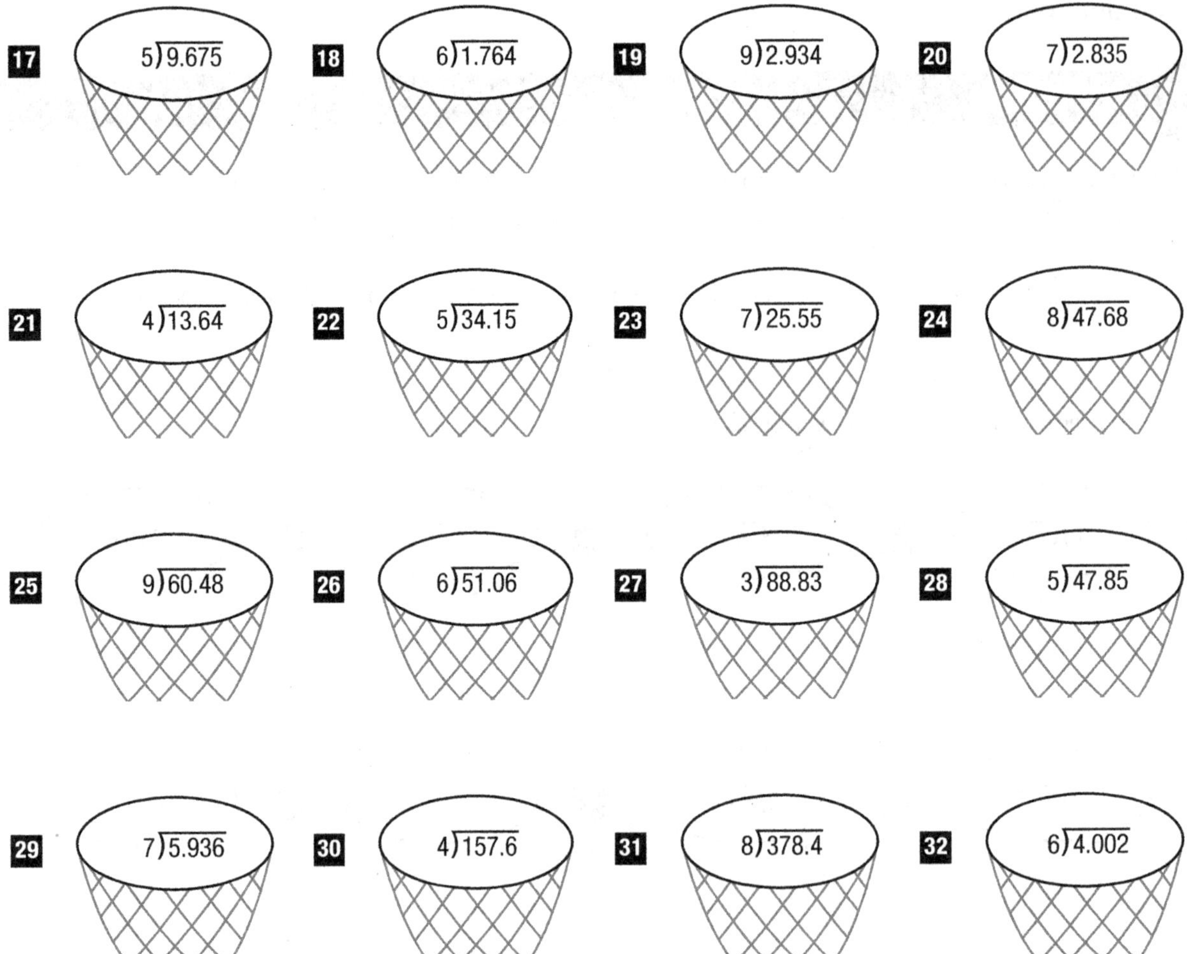

33 Complete this table.

	Number of items and cost per item	Cost	Change from K100
a	8 @ ________ each	K58.80	
b	6 @ ________ each	K69.90	
c	3 @ ________ each		K23.80
d	5 @ ________ each	K84.75	
e	2 @ ________ each		K26.40

	Number of items and cost per item	Cost	Change from K100
f	6 @ ________ each	K76.80	
g	5 @ ________ each		K32.50
h	3 @ ________ each		K6.25
i	7 @ ________ each	K39.55	
j	8 @ ________ each	K68.80	

Solve decimal word problems

Decide which process to use and then solve each word problem.

1 One metre of material costs K6.85. How much do the following lengths cost?

a 7 metres **b** 4 metres **c** 9 metres **d** 5 metres

2 Amos travelled 83.7 kilometres on Monday and 58.85 kilometres on Tuesday.

a How much further did Amos travel on Monday than Tuesday?
b What was the total distance travelled by Amos on the two days?

3 Four new tyres cost K314.60. What is the price of each tyre?

4 A school buys seven books at a cost of K8.55 each. What is the total cost?

5 A water tank had 745 litres of water in it. If 278.5 litres were used, how much was still in the tank?

6 Over five weeks, Tom earned K384.40. What was the average amount he earned per week?

7 Nancy is planning a holiday. She has to pay K560.80 for her airfare, K28.75 for her travel insurance and K170.50 for her accommodation. What is the total amount she has to pay?

8 A tourist paid a total of K731.75 for five nights accommodation. What was the cost per night?

9 Peter weighs 72.6 kilograms, his younger brother weighs 57.85 kilograms and his sister weighs 63.55 kilograms.

a How much more does Peter weigh than his brother?
b How much more does Peter weigh than his sister?
c What is the total mass of the three teenagers?

10 One suitcase weighs 23.085 kilograms and another weighs 27.65 kilograms.

a What is the difference in mass between the suitcases?
b If the suitcases are put on a trolley that can carry a total mass of 75 kilograms, how much less than that mass are they?

11 Ruth travels 57.92 kilometres each day. How far does she travel in:

a 5 days? **b** 8 days? **c** 6 days? **d** 9 days?

12 A bucket holds 9.28 litres of water. If I pour out 3.675 litres, how much will be left in the bucket?

13 A dressmaker buys 18.88 metres of material to make eight identical dresses. How much material will she use for each dress?

14 A café owner buys eight tins of biscuits, each with a mass of 1.675 kilograms. What is the total mass of the biscuit tins?

15 Peter's bank account had K861.85 after he withdrew K105.50 and K78.95. How much was in the account before he made the withdrawals?

6.1.3. Convert between simple fractions and decimals

Convert between fractions and decimals

1 Write the following fractions as decimals.

a $\frac{3}{10}$	**b** $\frac{27}{100}$	**c** $\frac{8}{10}$	**d** $\frac{51}{100}$	**e** $\frac{73}{100}$	**f** $\frac{1}{10}$
g $\frac{44}{100}$	**h** $\frac{5}{10}$	**i** $\frac{16}{100}$	**j** $\frac{81}{100}$	**k** $\frac{9}{10}$	**l** $\frac{39}{100}$
m $\frac{9}{100}$	**n** $\frac{5}{100}$	**o** $\frac{30}{100}$	**p** $\frac{90}{100}$	**q** $\frac{2}{100}$	**r** $\frac{10}{100}$
s $\frac{6}{10}$	**t** $\frac{72}{100}$	**u** $\frac{65}{100}$	**v** $\frac{7}{100}$	**w** $\frac{89}{100}$	**x** $\frac{2}{10}$

Remember

To convert a fraction that does not have a denominator of 10 or 100 to a decimal, first make an equivalent fraction with a denominator of 10 or 100.

Examples:

$\frac{4}{5} = \frac{8}{10} = 0.8$

$\frac{8}{25} = \frac{32}{100} = 0.32$

$\frac{6}{30} = \frac{2}{10} = 0.2$

2 Write the following fractions as decimals.

a $\frac{2}{5}$	**b** $\frac{13}{25}$	**c** $\frac{1}{4}$	**d** $\frac{3}{4}$
e $\frac{9}{50}$	**f** $\frac{1}{2}$	**g** $\frac{1}{20}$	**h** $\frac{3}{50}$
i $\frac{3}{5}$	**j** $\frac{6}{20}$	**k** $\frac{11}{25}$	**l** $\frac{1}{5}$
m $\frac{22}{25}$	**n** $\frac{15}{20}$	**o** $\frac{27}{50}$	**p** $\frac{2}{50}$
q $\frac{1}{25}$	**r** $\frac{13}{20}$	**s** $\frac{9}{60}$	**t** $\frac{20}{80}$
u $\frac{18}{75}$	**v** $\frac{16}{40}$	**w** $\frac{12}{30}$	**x** $\frac{54}{90}$

3 Write these mixed numbers as decimal numbers.

a $1\frac{3}{4}$ b $2\frac{1}{2}$ c $1\frac{7}{10}$ d $3\frac{11}{100}$ e $5\frac{1}{10}$ f $2\frac{1}{4}$

g $7\frac{29}{100}$ h $4\frac{2}{5}$ i $8\frac{3}{25}$ j $1\frac{4}{20}$ k $2\frac{9}{10}$ l $3\frac{1}{2}$

m $5\frac{1}{5}$ n $3\frac{22}{50}$ o $7\frac{1}{20}$ p $2\frac{9}{30}$ q $4\frac{21}{25}$ r $1\frac{3}{50}$

s $2\frac{24}{40}$ t $9\frac{33}{75}$ u $3\frac{9}{25}$ v $1\frac{27}{60}$ w $5\frac{18}{100}$ x $4\frac{5}{100}$

4 Write these improper fractions as decimal numbers.

a $\frac{22}{10}$ b $\frac{324}{100}$ c $\frac{17}{10}$ d $\frac{205}{100}$ e $\frac{183}{100}$ f $\frac{59}{10}$

g $\frac{21}{5}$ h $\frac{13}{4}$ i $\frac{9}{2}$ j $\frac{7}{5}$ k $\frac{27}{4}$ l $\frac{117}{50}$

m $\frac{81}{25}$ n $\frac{113}{20}$ o $\frac{39}{5}$ p $\frac{57}{25}$ q $\frac{31}{20}$ r $\frac{39}{4}$

s $\frac{25}{2}$ t $\frac{43}{5}$ u $\frac{128}{25}$ v $\frac{25}{4}$ w $\frac{53}{50}$ x $\frac{64}{5}$

5 Write the following decimals as fractions in their simplest form.

a 0.7 b 0.63 c 0.2 d 0.87

e 0.6 f 0.31 g 1.5 h 2.3

i 5.8 j 3.4 k 1.25 l 2.75

m 3.12 n 1.44 o 0.85 p 8.36

q 5.08 r 7.54 s 2.04 t 19.6

u 1.79 v 3.15 w 11.8 x 10.05

Remember

To convert a decimal to a fraction, first write it as a fraction out of 10 or 100 and then simplify it.

Examples.

$0.5 = \frac{5}{10} = \frac{1}{2}$

$0.75 = \frac{75}{100} = \frac{3}{4}$

$2.6 = 2\frac{6}{10} = 2\frac{3}{5}$

6 Put each group of fractions and decimals in order from smallest to largest.

a $\frac{7}{10}$ 0.43 $\frac{93}{100}$ b 0.8 $\frac{51}{100}$ 0.67 c $\frac{79}{100}$ $\frac{3}{10}$ 1.5

d $\frac{17}{10}$ 1.56 $1\frac{4}{10}$ e $\frac{4}{5}$ 0.9 $\frac{79}{100}$ f 0.13 $\frac{11}{100}$ 0.08

g $\frac{33}{50}$ $\frac{1}{2}$ 0.6 h $\frac{7}{20}$ $\frac{31}{100}$ 0.3 i $\frac{3}{4}$ 0.7 $\frac{12}{20}$

j 2.81 $2\frac{37}{100}$ $2\frac{4}{5}$ k 0.39 $\frac{9}{25}$ $\frac{3}{10}$ l 1.09 1.9 $1\frac{3}{5}$

m $\frac{1}{4}$ 0.2 $\frac{19}{100}$ n $\frac{41}{100}$ $\frac{22}{50}$ 0.45 o $\frac{12}{5}$ 2.37 $2\frac{11}{25}$

6.1.4. Link fractions and decimals to percentages and solve simple percentage problems

Relate fractions, decimals and percentages

Help Box

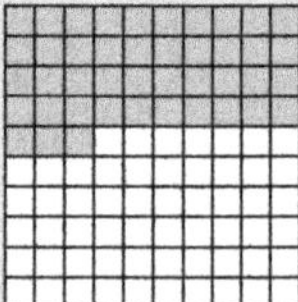

Percent means for every 100. Common and decimal fractions can be written as percentages. The symbol for percent is %.

This grid has 100 squares and 43 out of 100 are shaded. This means that 43 percent of the grid is shaded.

We can write this as $\frac{43}{100}$ or 0.43 or 43%.

Write a common fraction, a decimal fraction and a percentage for each grid.

1

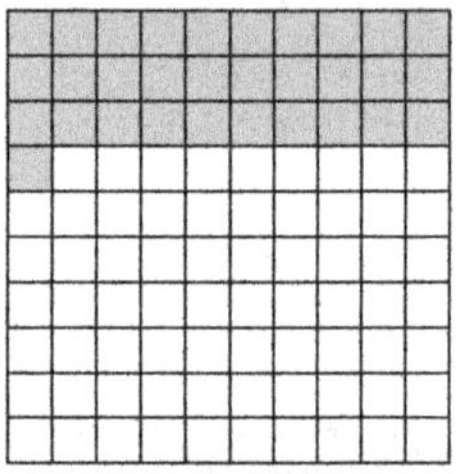

2

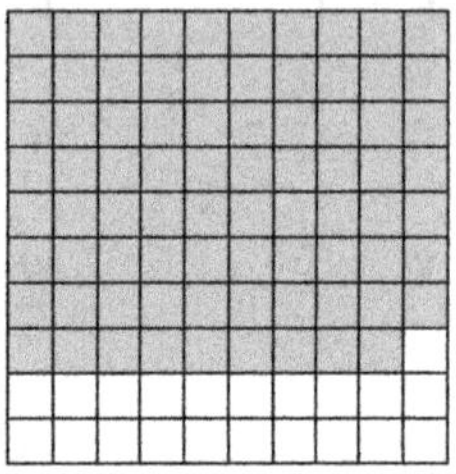

3

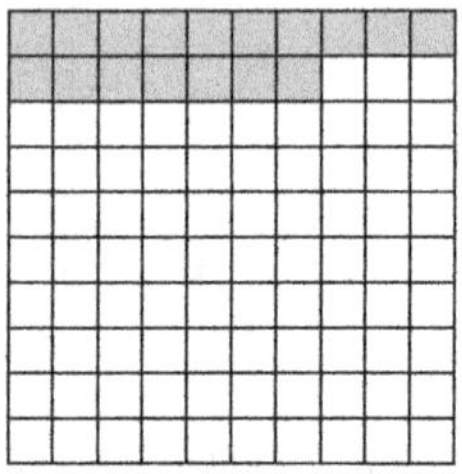

4

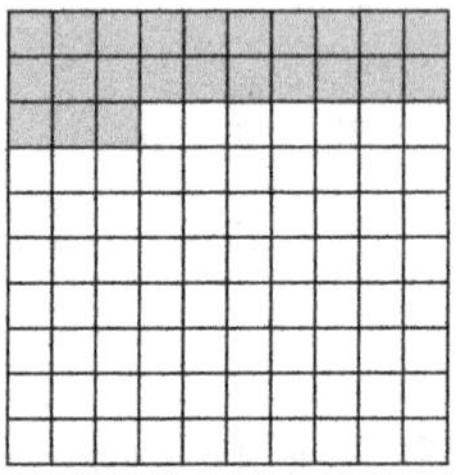

5

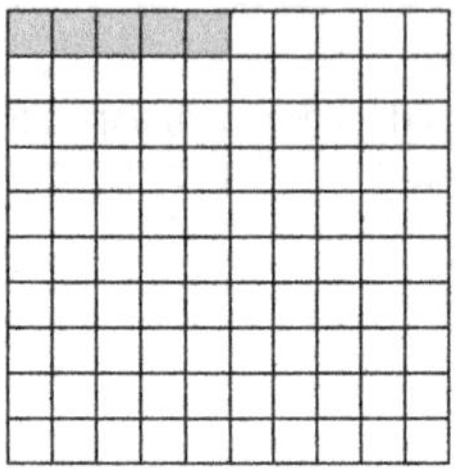

6

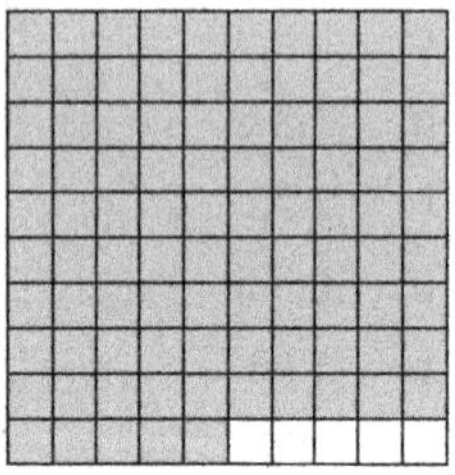

7

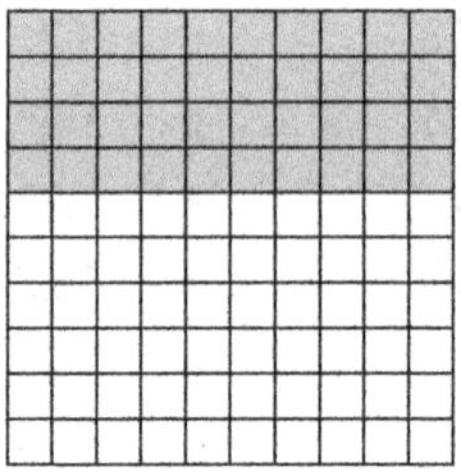

8

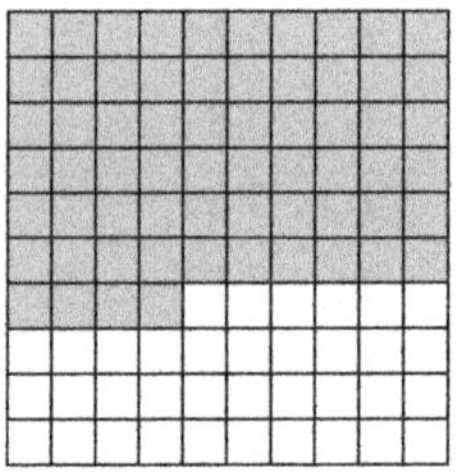

9

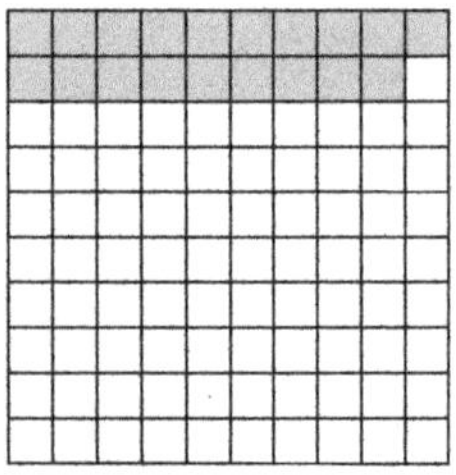

10

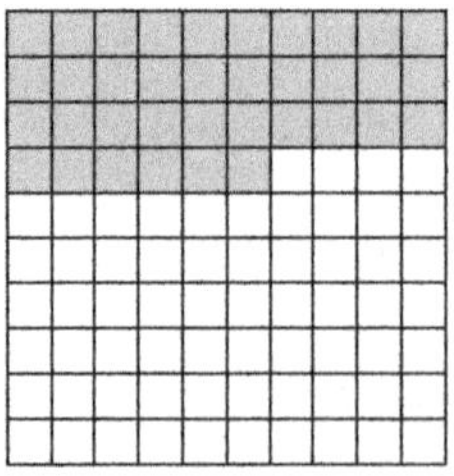

11

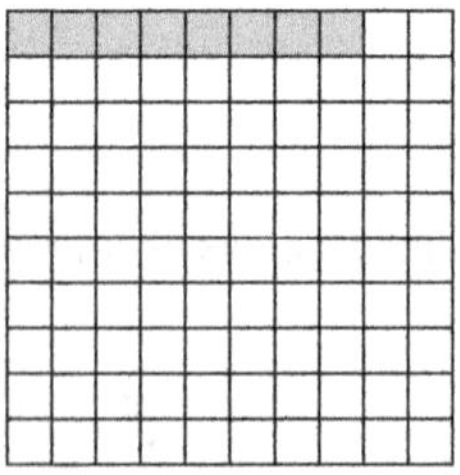

12

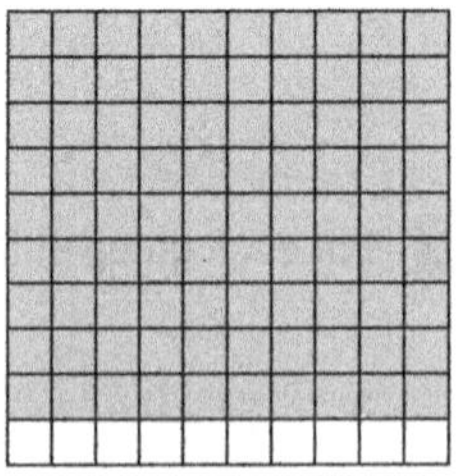

13

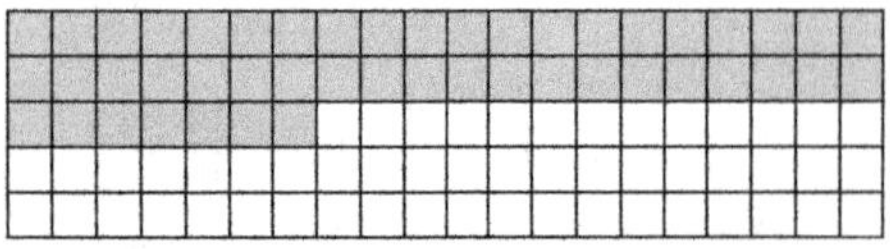

14

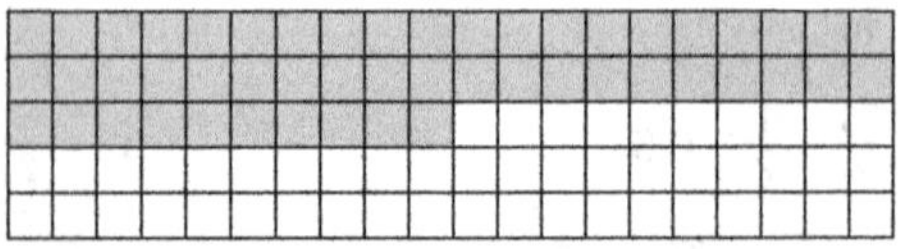

15

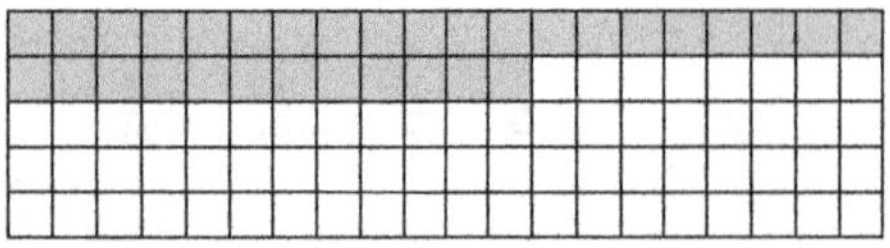

16

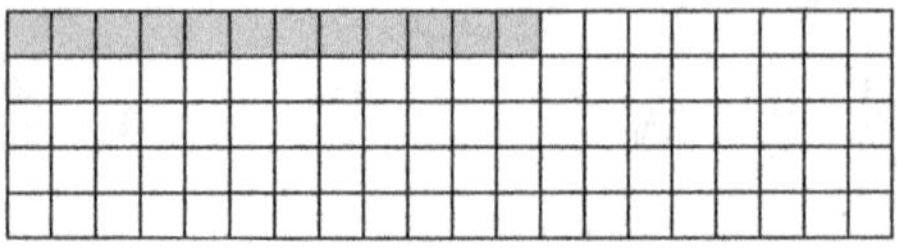

Convert between common fractions, decimals and percentages

1 Write these percentages as common fractions in their simplest form.

a	70%	**b**	81%	**c**	55%	**d**	30%	**e**	48%	**f**	15%
g	3%	**h**	26%	**i**	72%	**j**	64%	**k**	37%	**l**	8%
m	12%	**n**	90%	**o**	50%	**p**	41%	**q**	75%	**r**	67%
s	33%	**t**	59%	**u**	24%	**v**	99%	**w**	20%	**x**	10%

2 Write the common fractions from this group that:

a can be changed directly to percentages

b need to be written as fractions out of 100 first.

$\frac{37}{100}$ $\frac{15}{20}$ $\frac{7}{25}$ $\frac{4}{5}$ $\frac{8}{100}$ $\frac{33}{100}$ $\frac{9}{50}$

$\frac{9}{10}$ $\frac{46}{100}$ $\frac{2}{5}$ $\frac{11}{25}$ $\frac{1}{10}$ $\frac{79}{100}$ $\frac{8}{20}$

$\frac{85}{100}$ $\frac{3}{4}$ $\frac{4}{10}$ $\frac{17}{100}$ $\frac{28}{50}$ $\frac{20}{25}$ $\frac{61}{100}$

Remember

Common fractions with denominators of 100 can be changed directly to percentages:

$\frac{75}{100} = 75\%$

Common fractions with denominators other than 100 have to be written as fractions out of 100 first:

$\frac{7}{10} = \frac{70}{100} = 70\%$

3 Write these common fractions as percentages.

a	$\frac{39}{100}$	**b**	$\frac{21}{100}$	**c**	$\frac{45}{100}$	**d**	$\frac{13}{100}$	**e**	$\frac{67}{100}$	**f**	$\frac{7}{100}$
g	$\frac{1}{10}$	**h**	$\frac{9}{10}$	**i**	$\frac{2}{10}$	**j**	$\frac{34}{50}$	**k**	$\frac{16}{50}$	**l**	$\frac{40}{50}$
m	$\frac{6}{20}$	**n**	$\frac{3}{5}$	**o**	$\frac{6}{25}$	**p**	$\frac{9}{20}$	**q**	$\frac{21}{25}$	**r**	$\frac{1}{5}$
s	$\frac{11}{25}$	**t**	$\frac{17}{20}$	**u**	$\frac{3}{4}$	**v**	$\frac{1}{2}$	**w**	$\frac{1}{4}$	**x**	$\frac{5}{5}$

4 Write these percentages as decimals.

a	25%	**b**	10%	**c**	50%	**d**	75%	**e**	20%	**f**	17%
g	39%	**h**	95%	**i**	72%	**j**	61%	**k**	84%	**l**	42%
m	7%	**n**	4%	**o**	30%	**p**	48%	**q**	93%	**r**	11%
s	70%	**t**	35%	**u**	1%	**v**	22%	**w**	53%	**x**	9%

5 Write these decimals as percentages.

a	0.57	**b**	0.34	**c**	0.65	**d**	0.23	**e**	0.86	**f**	0.14
g	0.92	**h**	0.02	**i**	0.74	**j**	0.05	**k**	0.41	**l**	0.59
m	0.8	**n**	0.3	**o**	0.5	**p**	0.1	**q**	0.4	**r**	0.9
s	0.77	**t**	0.18	**u**	0.95	**v**	0.38	**w**	0.25	**x**	0.75

6 Copy and complete the charts to show how decimals, fractions and percentages relate to each other.

a

Fraction	$\frac{54}{100}$			$\frac{86}{100}$			$\frac{1}{2}$	
Decimal		0.39			0.7			0.31
Percentage			22%			5%		

b

Fraction		$\frac{7}{20}$			$\frac{1}{4}$			$\frac{2}{5}$
Decimal	0.08		0.5				0.55	
Percentage				75%		40%		

c

Fraction	$\frac{8}{25}$				$\frac{9}{10}$	$\frac{3}{50}$		
Decimal			0.03				0.91	0.2
Percentage		10%		100%				

d

Fraction			$\frac{1}{5}$				$\frac{19}{20}$	
Decimal		0.8				0.15		0.43
Percentage	67%			2%	16%			

e

Fraction		$\frac{27}{50}$			$\frac{47}{100}$		$\frac{16}{25}$	
Decimal	0.72			0.44				0.36
Percentage			30%			96%		

f

Fraction	$\frac{6}{10}$				$\frac{14}{20}$			
Decimal		0.13		0.09			0.33	
Percentage			57%			29%		87%

g

Fraction		$\frac{79}{100}$		$\frac{9}{20}$				$\frac{3}{4}$
Decimal			0.99			0.51		
Percentage	41%				23%		83%	

Compare fractions, decimals and percentages

1 Copy and complete the grid.

	%	Common fraction	Diagram	Number line	Decimal
a	25%			0 1	
b		$\frac{1}{5}$		0 1	
c				0 1	0.5
d		$\frac{3}{4}$		0 1	
e				0 1	
f	10%			0 1	

2 Write the largest common fraction, decimal or percentage in each group.

a 0.35 37% 3.5 $\frac{4}{10}$

b 23% 0.235 $\frac{22}{100}$ 0.2

c 0.57 0.507 55% $\frac{5}{10}$

d $\frac{7}{10}$ 0.07 $\frac{71}{100}$ 7%

e 0.215 26% $\frac{3}{10}$ $\frac{1}{4}$

f $\frac{1}{3}$ 0.35 30% $\frac{37}{100}$

g 16% 0.6 $\frac{1}{5}$ 1.6

h 95% 0.095 $\frac{9}{10}$ 0.93

i $\frac{47}{100}$ $\frac{1}{2}$ 0.45 $\frac{49}{100}$

j 3% 0.3 0.035 35%

k 0.72 $\frac{3}{4}$ $\frac{7}{10}$ 71%

l 45% 0.4 $\frac{5}{10}$ 0.04

m $\frac{57}{100}$ 0.56 $\frac{1}{2}$ 55%

n 0.88 80% 0.8 $\frac{85}{100}$

o $\frac{3}{5}$ 0.65 0.5 63%

p 41% 4.1 0.44 $\frac{45}{100}$

q $\frac{9}{10}$ 0.09 91% 0.93

r 0.15 $\frac{5}{20}$ 1.5 10%

s 3.6 36% 0.35 $\frac{33}{100}$

t 24% $\frac{15}{50}$ 0.2 $\frac{1}{4}$

3 Write >, < or = between each pair to show their relationship.

a 38% □ 0.38

b 59% □ 0.6

c $\frac{4}{10}$ □ 41%

d $\frac{74}{100}$ □ 75%

e 63% □ $\frac{6}{10}$

f $\frac{1}{5}$ □ 20%

g 80% □ $\frac{8}{10}$

h $\frac{3}{5}$ □ 55%

i 40% □ 0.4

j 0.17 □ $\frac{17}{50}$

k 9.5 □ 95%

l $\frac{6}{25}$ □ 20%

m $\frac{3}{4}$ □ 72%

n 0.56 □ 59%

o $\frac{2}{5}$ □ 0.49

p $\frac{78}{100}$ □ 0.8

Solve percentage calculations

Help Box

25% of 16

Method 1:

25% of 16 = $\frac{1}{4}$ of 16

$\frac{1}{4}$ of 16 = 4

Method 2:

$\frac{25}{100} \times \frac{16}{1}$

$= \frac{1}{4} \times \frac{16}{1}$

$= \frac{16}{4} = 4$

Remember

When finding a percentage of some numbers, the answer is best written as a decimal number.

Examples:

10% of 73 = $\frac{1}{10}$ of 73 = 7.3

50% of 25 = $\frac{1}{2}$ of 25 = 12.5

1 Solve these calculations.

a	50% of 20	**b**	75% of 60	**c**	20% of 45
d	70% of 80	**e**	10% of 30	**f**	40% of 90
g	100% of 55	**h**	5% of 100	**i**	50% of 500
j	90% of 70	**k**	80% of 120	**l**	25% of 32
m	20% of 15	**n**	75% of 24	**o**	10% of 260
p	30% of 90	**q**	25% of 28	**r**	100% of 10
s	120% of 10	**t**	5% of 60	**u**	40% of 150
v	15% of 40	**w**	75% of 48	**x**	90% of 180

2 Write the answers to these calculations as decimal numbers.

a	10% of 57	**b**	50% of 19	**c**	25% of 21
d	20% of 18	**e**	75% of 17	**f**	10% of 86
g	20% of 72	**h**	50% of 85	**i**	5% of 42
j	25% of 63	**k**	75% of 29	**l**	10% of 64

3 Copy this table and calculate the amount of discount on each price.

	Cost	Percentage discount	Amount of discount
a	K50	10%	
b	K20	50%	
c	K60	25%	
d	K10	10%	
e	K10	15%	
f	K30	30%	
g	K150	20%	
h	K40	5%	
i	K50	15%	
j	K160	20%	

4 Copy this table and calculate the price after discount.

	Cost	Percentage discount	Price after discount
a	K25	20%	
b	K100	5%	
c	K60	10%	
d	K500	25%	
e	K150	30%	
f	K45	10%	
g	K28	5%	
h	K70	25%	
i	K36	20%	
j	K120	15%	

6.1.8 Use indices to the power of 2 and 3

Use indices in calculations

Remember

Multiplication facts can be written using indices. Examples: $5 \times 5 = 5^2$ $4 \times 4 \times 4 = 4^3$

1 Write each of these as multiplications.

a 3^2 b 5^3 c 4^2 d 1^3 e 9^2 f 6^3
g 8^3 h 10^3 i 7^2 j 12^2 k 15^3 l 11^2
m 5^2 n 2^2 o 3^3 p 9^3 q 10^2 r 4^3

2 Use index notation to write these multiplications.

a $9 \times 9 \times 9$ b $7 \times 7 \times 7$ c 6×6 d $3 \times 3 \times 3$ e 13×13 f 2×2
g 10×10 h 18×18 i $8 \times 8 \times 8$ j $2 \times 2 \times 2$ k 14×14 l 21×21
m 3×3 n $1 \times 1 \times 1$ o $5 \times 5 \times 5$ p 11×11 q 9×9 r 16×16

3 Calculate answers to these.

a 4^2 b 5^2 c 10^2 d 8^2 e 15^2 f 20^2
g 2^3 h 1^3 i 5^3 j 10^3 k 4^3 l 7^3
m 9^2 n 6^3 o 12^2 p 9^3 q 30^2 r 20^3

4 Calculate the answers to the following multiplication problems.

a $3^2 \times 6$ b $2^3 \times 8$ c $10^3 \times 2$ d $5^2 \times 8$ e $5^3 \times 2$ f $8^2 \times 7$
g $1^3 \times 9$ h $9^2 \times 4$ i $6^2 \times 3$ j $4^3 \times 5$ k $3^2 \times 8$ l $10^3 \times 6$
m $8^2 \times 2$ n $3^3 \times 3$ o $2^3 \times 5$ p $7^2 \times 10$ q $6^3 \times 2$ r $12^2 \times 3$

5 Calculate the answers to these additions and subtractions.

a $10^2 - 9^2$ b $8^2 + 2^2$ c $2^3 + 10^3$ d $5^3 - 7^2$ e $4^2 + 6^2$
f $5^2 - 1^3$ g $3^3 - 3^2$ h $10^2 + 5^3$ i $3^2 + 1^2 + 5^2$ j $4^3 - 8^2$
k $6^2 + 6^2 + 2^3$ l $6^3 - 9^2$ m $2^2 + 1^3 + 7^2$ n $10^3 - 10^2$ o $3^3 + 5^3 + 6^2$

6 Use =, > or < to complete the following problems.

a 6^2 ☐ 40 b 4^2 ☐ 15 c 20 ☐ 5^2 d 3^2 ☐ 8 e 50 ☐ 7^2
f 60 ☐ 8^2 g 45 ☐ 7^2 h 36 ☐ 6^2 i 95 ☐ 9^2 j 6^2 ☐ 35
k 10^2 ☐ 100 l 5^2 ☐ 27 m 2^3 ☐ 12 n 45 ☐ 3^3 o 500 ☐ 10^3

Assessment Number and Application

Multiple choice fraction test

1 Which one of these fractions is equivalent to $\frac{2}{5}$?

a $\frac{9}{10}$ b $\frac{6}{9}$ c $\frac{6}{15}$ d $\frac{1}{3}$

2 Which one of these fractions is equivalent to $\frac{3}{12}$?

a $\frac{6}{20}$ b $\frac{1}{4}$ c $\frac{1}{3}$ d $\frac{9}{24}$

3 The simplest form of $\frac{8}{24}$ is:

a $\frac{1}{6}$ b $\frac{1}{4}$ c $\frac{1}{3}$ d $\frac{2}{6}$

4 The simplest form of $\frac{12}{32}$ is:

a $\frac{6}{16}$ b $\frac{3}{8}$ c $\frac{4}{8}$ d $\frac{2}{5}$

5 Which of these is an improper fraction?

a $\frac{7}{3}$ b $\frac{5}{9}$ c $1\frac{1}{3}$ d $\frac{2}{5}$

6 What is $\frac{13}{4}$ as a mixed number?

a $4\frac{1}{4}$ b $3\frac{3}{4}$ c $1\frac{3}{4}$ d $3\frac{1}{4}$

7 What is $6\frac{3}{5}$ as an improper fraction?

a $\frac{63}{5}$ b $\frac{30}{5}$ c $\frac{33}{5}$ d $\frac{21}{5}$

8 Which fraction is larger than $3\frac{5}{8}$?

a $\frac{25}{8}$ b $\frac{31}{8}$ c $3\frac{1}{4}$ d $3\frac{1}{2}$

9 Which fraction is smaller than $\frac{23}{9}$?

a $2\frac{2}{3}$ b $2\frac{8}{9}$ c $2\frac{1}{3}$ d $2\frac{7}{9}$

10 What is the answer to $\frac{5}{6} + \frac{5}{6}$?

a $\frac{10}{12}$ b $1\frac{5}{6}$ c $1\frac{2}{3}$ d $1\frac{2}{10}$

11 What is the answer to $\frac{3}{4} + \frac{5}{12}$?

a $1\frac{1}{6}$ b $\frac{8}{16}$ c $\frac{8}{12}$ d $1\frac{8}{12}$

12 What is the answer to $\frac{7}{8} - \frac{2}{5}$?

a $\frac{5}{3}$ b $\frac{19}{40}$ c $1\frac{2}{3}$ d $\frac{1}{2}$

13 What is the answer to $3\frac{3}{8} - 1\frac{3}{4}$?

a $2\frac{3}{4}$ b $1\frac{3}{8}$ c $2\frac{1}{4}$ d $1\frac{5}{8}$

14 What is the answer to $7 \times \frac{5}{6}$?

a $\frac{35}{42}$ b $5\frac{5}{6}$ c $\frac{5}{42}$ d $7\frac{5}{6}$

15 $\frac{2}{9}$ of 72 is:

a 16 b 20 c 8 d 10

16 What is the answer to $\frac{6}{7} \times \frac{2}{3}$ in its simplest form?

a $\frac{8}{10}$ b $\frac{12}{21}$ c $\frac{14}{18}$ d $\frac{4}{7}$

17 What is the answer to $1\frac{5}{8} \times 2\frac{2}{3}$?

a $2\frac{5}{12}$ b $3\frac{7}{11}$ c $4\frac{1}{3}$ d $4\frac{1}{4}$

18 What is the answer to $\frac{2}{3} \div 8$?

a $\frac{1}{8}$ b $5\frac{1}{3}$ c $\frac{1}{24}$ d $\frac{1}{12}$

19 What is the answer to $\frac{5}{9} \div \frac{1}{5}$?

a $\frac{5}{45}$ b $2\frac{7}{9}$ c $\frac{5}{14}$ d $\frac{25}{10}$

20 What is the answer to $4\frac{1}{5} \div 1\frac{3}{4}$?

a $1\frac{1}{2}$ b $2\frac{2}{5}$ c $7\frac{7}{20}$ d $1\frac{5}{7}$

Assessment Number and Application

1 Solve these decimal additions and subtractions.

a	12.78 + 29.85	**b**	56.09 + 37.66	**c**	9.685 + 15.57	**d**	38.49 + 153.63
e	72.38 – 49.67	**f**	51.04 – 27.25	**g**	63.4 – 39.625	**h**	84.16 – 35.9

2 Solve these decimal multiplications.

a	6.5 × 7	**b**	8.3 × 9	**c**	5.7 × 6	**d**	9.6 × 4
e	23.8 × 3	**f**	19.4 × 8	**g**	36.9 × 5	**h**	45.2 × 7
i	6.83 × 4	**j**	4.96 × 6	**k**	11.55 × 9	**l**	36.49 × 8

3 Solve these decimal divisions.

a	$2\overline{)6.4}$	**b**	$3\overline{)6.9}$	**c**	$4\overline{)44.8}$	**d**	$3\overline{)9.69}$
e	$7\overline{)9.1}$	**f**	$6\overline{)8.4}$	**g**	$8\overline{)35.2}$	**h**	$5\overline{)27.5}$
i	$4\overline{)1.836}$	**j**	$8\overline{)5.728}$	**k**	$6\overline{)53.22}$	**l**	$7\overline{)41.16}$

4 Write these fractions as decimals.

a	$\frac{7}{10}$	**b**	$\frac{59}{100}$	**c**	$\frac{34}{100}$	**d**	$\frac{1}{10}$	**e**	$\frac{4}{5}$	**f**	$\frac{17}{25}$
g	$\frac{9}{50}$	**h**	$\frac{23}{50}$	**i**	$\frac{7}{20}$	**j**	$\frac{22}{40}$	**k**	$\frac{24}{80}$	**l**	$\frac{30}{75}$

5 Write these decimals as fractions in their simplest form.

a	0.9	**b**	0.81	**c**	0.8	**d**	0.24	**e**	0.56	**f**	0.42
g	2.5	**h**	4.75	**i**	1.06	**j**	9.6	**k**	11.25	**l**	4.15

6 Write these percentages as common fractions in their simplest form.

a	80%	**b**	25%	**c**	10%	**d**	38%	**e**	51%	**f**	16%
g	63%	**h**	96%	**i**	4%	**j**	74%	**k**	22%	**l**	45%

Assessment Number and Application

7 Write these common fractions as percentages.

a	$\frac{31}{100}$	**b**	$\frac{6}{100}$	**c**	$\frac{9}{50}$	**d**	$\frac{27}{50}$	**e**	$\frac{7}{20}$	**f**	$\frac{9}{25}$
g	$\frac{9}{10}$	**h**	$\frac{2}{5}$	**i**	$\frac{3}{4}$	**j**	$\frac{16}{20}$	**k**	$\frac{22}{25}$	**l**	$\frac{18}{20}$

8 Write these percentages as decimals.

a	98%	**b**	3%	**c**	13%	**d**	58%	**e**	41%	**f**	76%
g	11%	**h**	29%	**i**	62%	**j**	35%	**k**	84%	**l**	90%

9 Write these decimals as percentages.

a	0.61	**b**	0.4	**c**	0.07	**d**	0.55	**e**	0.89	**f**	0.36
g	0.2	**h**	0.97	**i**	0.14	**j**	0.76	**k**	0.23	**l**	0.42

10 Which is larger?

a	49% or 0.4	**b**	72% or $\frac{7}{10}$	**c**	$\frac{2}{5}$ or 25%	**d**	0.2 or $\frac{25}{100}$
e	0.75 or 80%	**f**	0.05 or $\frac{5}{10}$	**g**	10% or 0.01	**h**	1.0 or $\frac{1}{10}$
i	$\frac{17}{20}$ or 0.8	**j**	63% or 6.3	**k**	$\frac{14}{25}$ or 50%	**l**	0.11 or $\frac{11}{20}$

11 Put each set of common fractions, decimals and percentages in order from largest to smallest.

a	0.7, $\frac{75}{100}$, 7%	**b**	$\frac{5}{20}$, 20%, 0.05	**c**	37%, 0.3, $\frac{4}{10}$	**d**	46%, 0.046, $\frac{4}{100}$
e	0.85, $\frac{9}{10}$, 87%	**f**	$\frac{1}{5}$, 17%, 0.7	**g**	0.09, $\frac{3}{50}$, 90%	**h**	$\frac{1}{3}$, 36%, 0.35
i	0.65, $\frac{3}{5}$, 63%	**j**	5%, 0.55, $\frac{26}{50}$	**k**	90%, $\frac{19}{20}$, 0.92	**l**	$\frac{1}{4}$, 0.02, 20%

12 Solve these percent calculations.

a	25% of 64	**b**	50% of 72	**c**	75% of 36	**d**	10% of 170
e	20% of 85	**f**	40% of 60	**g**	5% of 40	**h**	30% of 50
i	70% of 80	**j**	90% of 100	**k**	15% of 20	**l**	60% of 90

Assessment Number and Application

13 Write the answers to these calculations as decimal numbers.

a	50% of 21	**b**	25% of 9	**c**	10% of 87	**d**	20% of 38
e	75% of 33	**f**	5% of 26	**g**	50% of 55	**h**	25% of 45
i	10% of 72	**j**	20% of 63	**k**	75% of 49	**l**	5% of 58

14 Write as multiplications.

a	20^2	**b**	4^3	**c**	8^2	**d**	15^3	**e**	25^2	**f**	12^3
g	18^3	**h**	50^2	**i**	24^3	**j**	54^2	**k**	75^3	**l**	63^2

15 Use index notation to record these multiplications.

a	14×14	**b**	30×30	**c**	$19 \times 19 \times 19$	**d**	$11 \times 11 \times 11$
e	26×26	**f**	$45 \times 45 \times 45$	**g**	17×17	**h**	$32 \times 32 \times 32$

16 Calculate answers to these.

a	$4^2 \times 3$	**b**	$2^3 \times 6$	**c**	$5^2 \times 10$	**d**	$10^3 \times 2$
e	$11^2 - 7^2$	**f**	$4^3 + 5^2$	**g**	$20^2 - 3^3$	**h**	$6^3 + 8^2$
i	$6^2 \times 2^2$	**j**	$5^3 \times 2^3$	**k**	$12^2 - 4^3$	**l**	$10^3 + 7^3$

17 Calculate answers to these problems.

a Find the product of 25 364 and 18.

b What is the total of 7066, 32 784 and 20 358?

c Find the difference between 51 231 and 28 655.

d What is the quotient of 45 309 and 33?

e Find the sum of 18 989, 45 067 and 25 625.

f What is the product of 38 094 and 9?

Strand	Space and Shape

6.2.2. Convert between metric units of length and round off

Convert between metric units of length

1 Write these metre (m) lengths as centimetres (cm).

a	7 m	**b**	3 m	**c**	$5\frac{1}{2}$ m	**d**	$8\frac{1}{2}$ m	**e**	$3\frac{1}{4}$ m	**f**	$4\frac{3}{4}$ m
g	$3\frac{1}{10}$ m	**h**	$1\frac{9}{10}$ m	**i**	$6\frac{3}{10}$ m	**j**	$1\frac{4}{5}$ m	**k**	$2\frac{1}{5}$ m	**l**	$\frac{7}{10}$ m

2 Write these centimetre (cm) lengths as millimetres (mm).

a	5 cm	**b**	$3\frac{1}{2}$ cm	**c**	10 cm	**d**	$2\frac{1}{2}$ cm	**e**	12 cm	**f**	30 cm
g	15 cm	**h**	$8\frac{1}{2}$ cm	**i**	25 cm	**j**	36 cm	**k**	$18\frac{1}{2}$ cm	**l**	100 cm

3 Write these millimetre (mm) lengths as centimetres (cm).

a	30 mm	**b**	70 mm	**c**	10 mm	**d**	150 mm	**e**	110 mm	**f**	90 mm
g	200 mm	**h**	180 mm	**i**	500 mm	**j**	350 mm	**k**	480 mm	**l**	1000 mm

4 Write these centimetre (cm) lengths as metres (m).

a	500 cm	**b**	1200 cm	**c**	800 cm	**d**	2000 cm	**e**	3600 cm	**f**	1400 cm
g	5900 cm	**h**	4100 cm	**i**	8500 cm	**j**	10 000 cm	**k**	12 000 cm	**l**	11 600 cm

5 Write these metre (m) lengths as millimetres (mm).

a	7 m	**b**	12 m	**c**	8 m	**d**	2 m	**e**	$4\frac{1}{2}$ m	**f**	$10\frac{1}{2}$ m
g	$1\frac{1}{4}$ m	**h**	$5\frac{1}{4}$ m	**i**	$3\frac{3}{4}$ m	**j**	$6\frac{3}{4}$ m	**k**	$\frac{9}{10}$ m	**l**	$11\frac{7}{10}$ m

6 Write these millimetre (mm) lengths as metres (m).

a	3000 mm	**b**	5000 mm	**c**	14 000 mm	**d**	11 000 mm
e	6500 mm	**f**	1500 mm	**g**	12 500 mm	**h**	1250 mm
i	7250 mm	**j**	9750 mm	**k**	750 mm	**l**	250 mm

7 Write these kilometre (km) lengths as metres (m).

a	5 km	b	$10\frac{1}{2}$ km	c	$4\frac{1}{4}$ km	d	$7\frac{3}{4}$ km	e	$\frac{1}{5}$ km	f	$\frac{9}{10}$ km
g	$1\frac{3}{10}$ km	h	25 km	i	$11\frac{1}{4}$ km	j	$8\frac{4}{5}$ km	k	$6\frac{7}{10}$ km	l	$12\frac{1}{8}$ km

8 Write these metre (m) lengths as kilometres (km).

a	4000 m	b	9000 m	c	15 000 m	d	22 000 m
e	2500 m	f	11 500 m	g	40 500 m	h	250 m
i	5250 m	j	7750 m	k	16 750 m	l	13 200 m

9 Write these metre (m) lengths as centimetres (cm).

a	2.06 m	b	1.4 m	c	5.45 m	d	0.7 m	e	3.18 m	f	4.8 m
g	1.07 m	h	6.65 m	i	13.8 m	j	15.02 m	k	11.5 m	l	10.48 m

10 Write these centimetre (cm) lengths as a decimal fraction of a metre (m).

a	243 cm	b	167 cm	c	79 cm	d	316 cm	e	540 cm	f	112 cm
g	33 cm	h	289 cm	i	804 cm	j	950 cm	k	311 cm	l	765 cm

11 Write these centimetre (cm) lengths as millimetres (mm).

a	2.4 cm	b	4.7 cm	c	1.3 cm	d	8.5 cm	e	3.8 cm	f	11.2 cm
g	5.6 cm	h	12.9 cm	i	0.8 cm	j	10.5 cm	k	0.1 cm	l	15.8 cm

12 Write these millimetre (mm) lengths as a decimal fraction of a centimetre (cm).

a	129 mm	b	413 mm	c	305 mm	d	78 mm	e	565 mm	f	53 mm
g	237 mm	h	41 mm	i	472 mm	j	134 mm	k	260 mm	l	80 mm

13 Write these kilometre (km) lengths as metres (m).

a	1.8 km	b	4.6 km	c	7.9 km	d	15.1 km	e	0.3 km	f	3.25 km
g	8.37 km	h	5.02 km	i	11.79 km	j	1.672 km	k	6.015 km	l	0.916 km

14 Write these metre (m) lengths as a decimal fraction of a kilometre (km).

a	2800 m	b	5900 m	c	8700 m	d	3450 m	e	1375 m	f	1059 m
g	925 m	h	709 m	i	10 625 m	j	12 004 m	k	18 300 m	l	11 610 m

15 Write these metre (m) lengths as millimetres (mm).

a	1.5 m	**b**	3.9 m	**c**	0.4 m	**d**	9.7 m	**e**	12.6 m	**f**	10.3 m
g	6.25 m	**h**	4.08 m	**i**	7.75 m	**j**	11.02 m	**k**	4.176 m	**l**	0.978 m

16 Write these millimetre (mm) lengths as a decimal fraction of a metre (m).

a	3100 mm	**b**	2800 mm	**c**	6300 mm	**d**	9850 mm
e	5076 mm	**f**	8109 mm	**g**	635 mm	**h**	884 mm
i	17 865 mm	**j**	23 200 mm	**k**	31 090 mm	**l**	12 008 mm

Add mixed units of length and round off

1 Convert the different lengths to the same unit of measurement and calculate.

a	400 cm + 3.7 m + 150 mm	**b**	5.85 m + 0.5 km + 1200 cm
c	1.4 km + 560 m + 0.865 km	**d**	0.9 m + 6500 mm + 1.07 m
e	7250 mm + 1550 cm + 24 m	**f**	3.8 km + 5000 cm + 2230 m
g	$2\frac{4}{5}$ m + 0.75 m + 380 cm	**h**	$\frac{1}{2}$ cm + 65 mm + 1.9 cm
i	$12\frac{3}{4}$ km + 6250 m + 8.5 km	**j**	10.5 m + 375 cm + $2\frac{7}{10}$ m
k	0.2 km + 85 m + 340 cm	**l**	20 mm + $\frac{3}{10}$ cm + 8.6 cm
m	9.08 m + $\frac{3}{5}$ m + 400 mm	**n**	$17\frac{1}{2}$ cm + 580 mm + 1.3 m
o	$3\frac{9}{10}$ m + 1500 cm + 0.7 m	**p**	2.85 km + 14 600 cm + 1765 m
q	34.9 cm + 118 mm + $17\frac{2}{5}$ cm	**r**	250 cm + 5.1 m + 840 mm
s	6.9 km + 20 000 cm + 830 m	**t**	$2\frac{3}{4}$ km + 1630 m + 1.1 km

2 Round these lengths to the nearest metre (m).

a	57 cm	**b**	617 cm	**c**	368 cm	**d**	870 mm	**e**	2300 mm
f	2.75 m	**g**	15.4 m	**h**	1684 cm	**i**	3119 cm	**j**	10.06 m
k	1615 mm	**l**	782 cm	**m**	9.05 m	**n**	1690 mm	**o**	7.27 m
p	3.81 m	**q**	814 cm	**r**	4212 mm	**s**	27.39 m	**t**	2.5 m

3 Round these lengths to the nearest centimetre (cm).

a	39 mm	**b**	1.67 cm	**c**	516 mm	**d**	32.5 cm	**e**	4.8 cm
f	855 mm	**g**	51 mm	**h**	18.5 cm	**i**	783 mm	**j**	1024 mm
k	13.05 cm	**l**	79 mm	**m**	115 mm	**n**	22 mm	**o**	57.8 cm
p	208 mm	**q**	10.7 cm	**r**	652 mm	**s**	7166 mm	**t**	5025 mm

4 Round these lengths to the nearest kilometre (km).

a	9.05 km	b	15.8 km	c	817 m	d	4563 m	e	0.566 km
f	7.25 km	g	7112 m	h	10 967 m	i	27.5 km	j	18.065 km
k	31 714 m	l	23.632 km	m	1008 m	n	3600 m	o	9190 m
p	5.575 km	q	12 235 m	r	2.16 km	s	594 m	t	20 070 m

5 From these lengths, record those that can be rounded off to 7 metres.

a	674 cm	b	7.28 m	c	607 cm	d	7390 mm	e	6.15 m
f	7.49 m	g	650 cm	h	6600 mm	i	7.95 m	j	7650 mm

6 From these lengths, record those that can be rounded off to 10 kilometres.

a	9500 m	b	10.8 km	c	10 400 m	d	9150 m	e	10 500 m
f	9.53 km	g	9499 m	h	10.499 km	i	10.51 km	j	9509 m

7 From these lengths, record those that can be rounded off to 6 centimetres.

a	45 mm	b	59 mm	c	6.8 cm	d	5.5 cm	e	64 mm
f	70 mm	g	0.6 cm	h	56 mm	i	0.065 m	j	0.058 m

8 Complete this table.

	Length	Nearest metre	Nearest kilometre
a	725.4 m		
b	908.31 m		
c	655.9 m		
d	1843.7 m		
e	3065.2 m		

	Length	Nearest metre	Nearest kilometre
f	2689.6 m		
g	1516.78 m		
h	5918.49 m		
i	8235.8 m		
j	4704.63 m		

9 Complete this table.

	Length	Nearest centimetre	Nearest metre
a	587 mm		
b	868 mm		
c	635 mm		
d	1588 mm		
e	1925 mm		

	Length	Nearest centimetre	Nearest metre
f	3154 mm		
g	5936 mm		
h	3781 mm		
i	8158 mm		
j	4712 mm		

6.2.3 Calculate the perimeter of shapes

Calculate the perimeter of various shapes

1 Use the measurements to calculate the perimeter of each plantation.

a

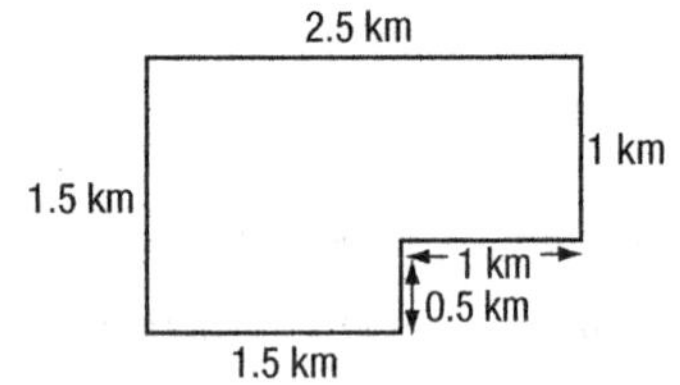

b

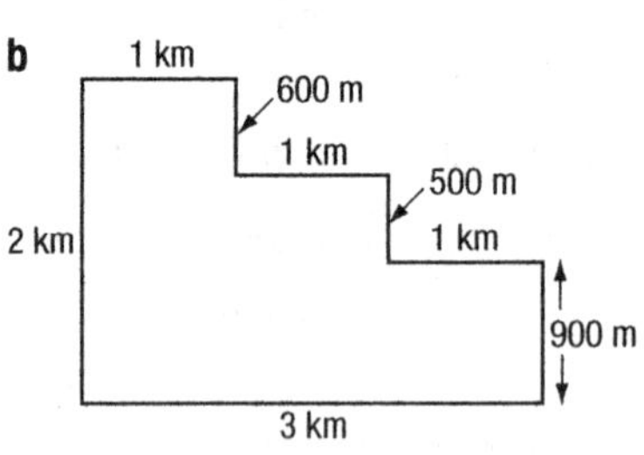

c

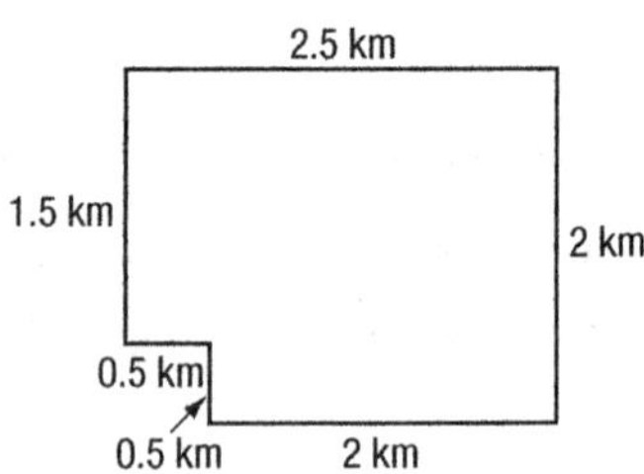

d

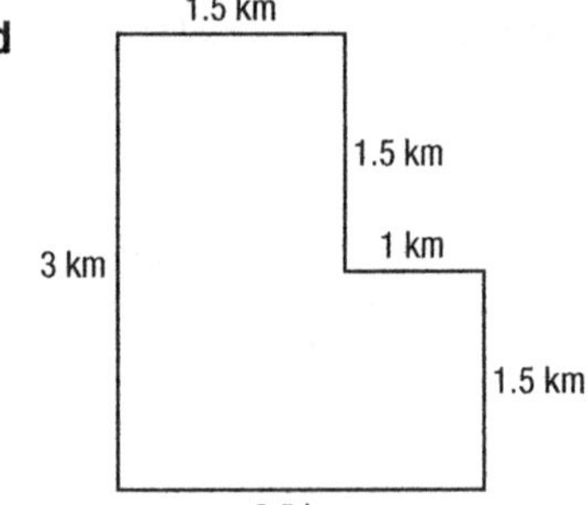

e

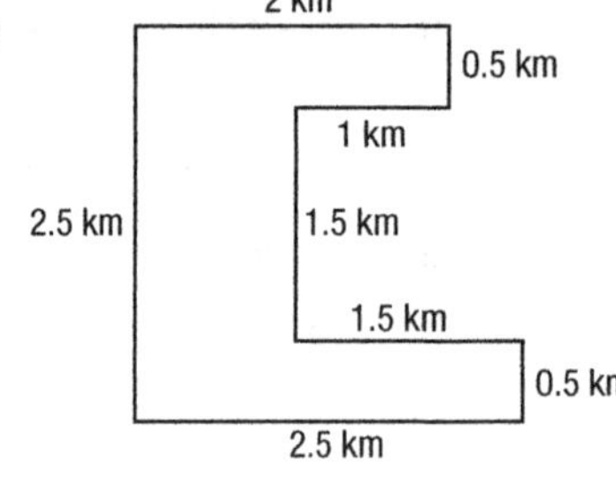

f

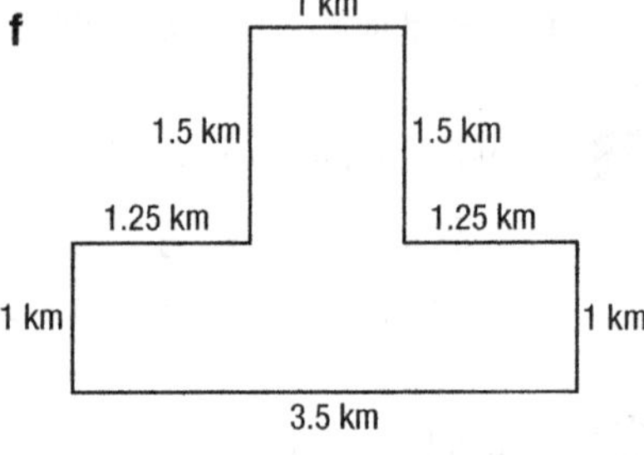

2 Use a ruler to help you find the perimeter of each shape.

a

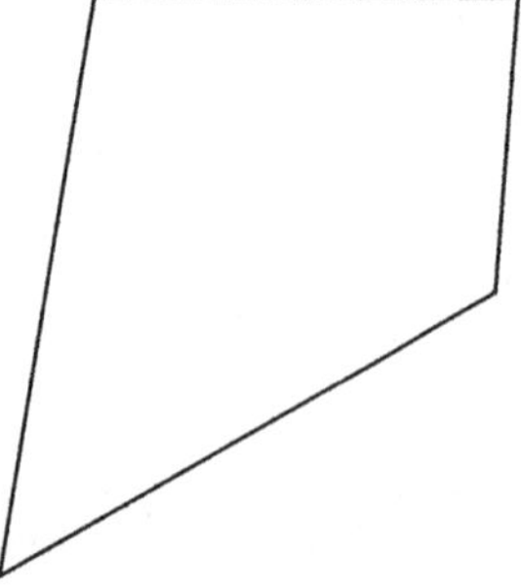

b

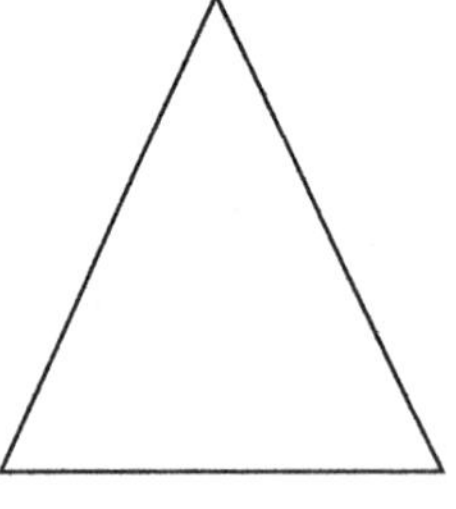

c

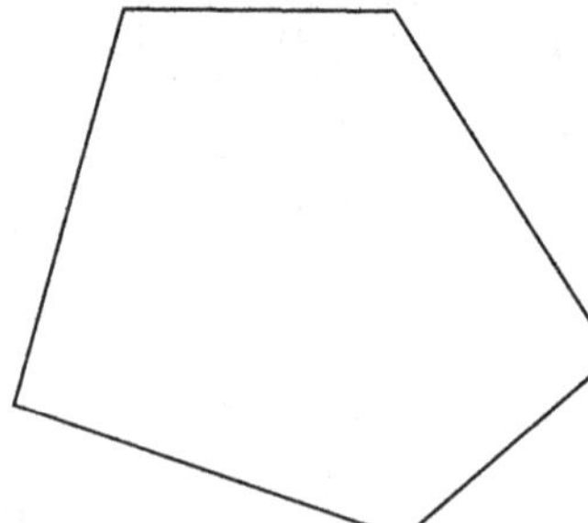

d

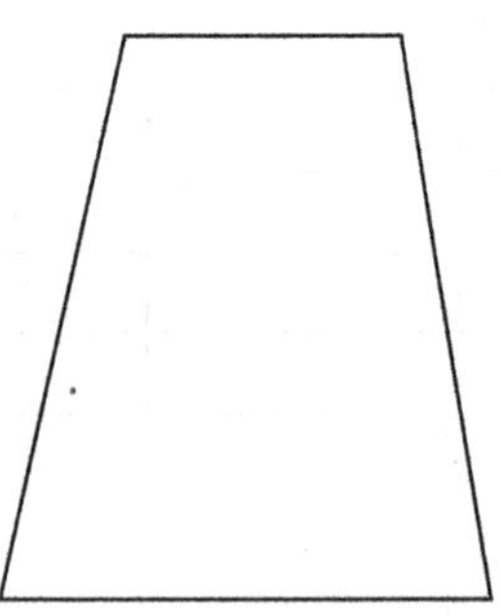

e

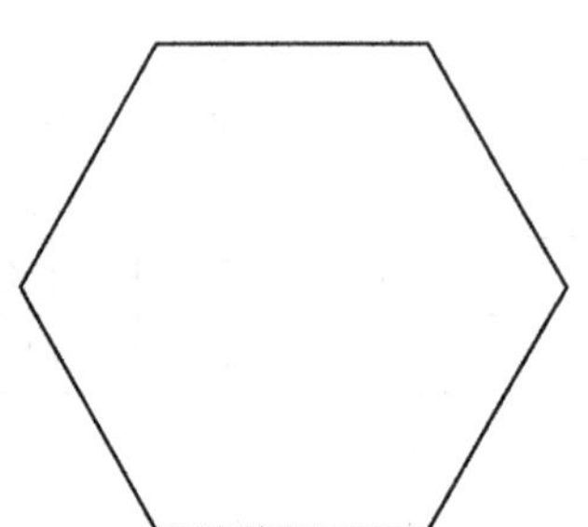

f

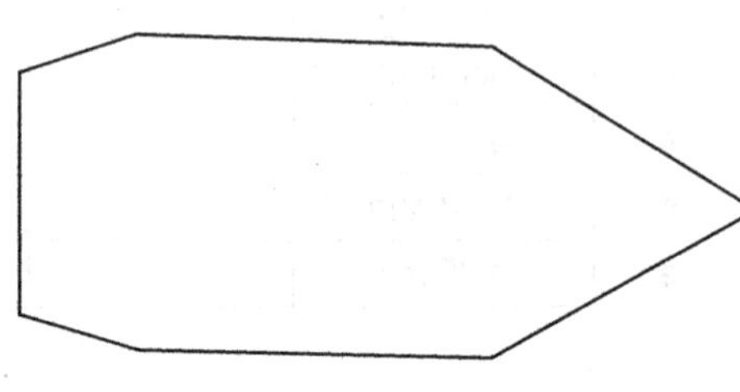

Calculate the perimeter of rectangles

1 Use the measurements to calculate the perimeter of each rectangle.

a

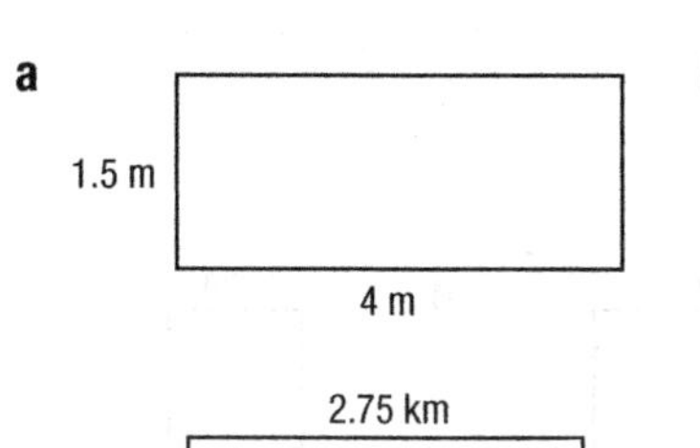

b

2 m

3m

c

d

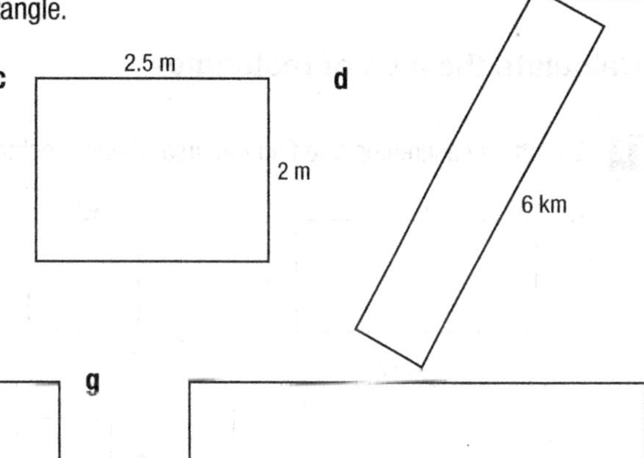

e

2.75 km

4.5 km

f

2 m

4.25 m

g

2.5 km

8 km

h

15 m

2 m

2 Measure the length and the width to help you calculate the perimeter of each rectangle.

a

b

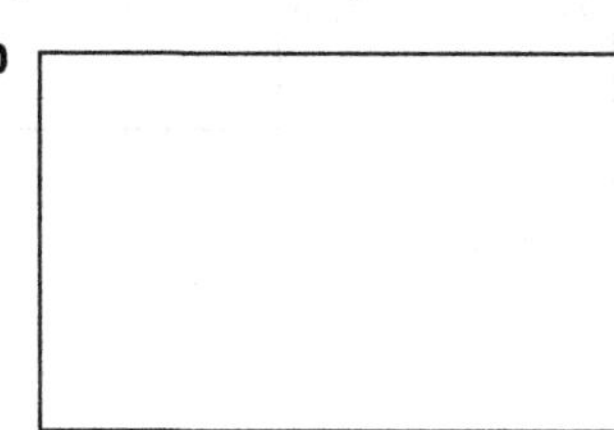

c

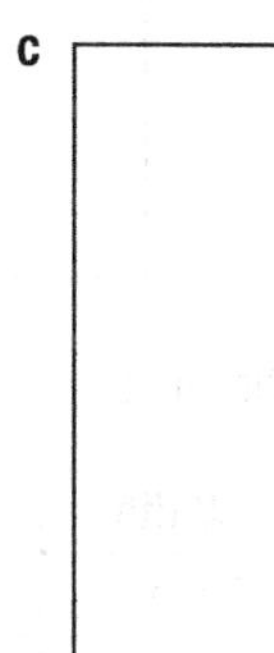

d

e

f

g

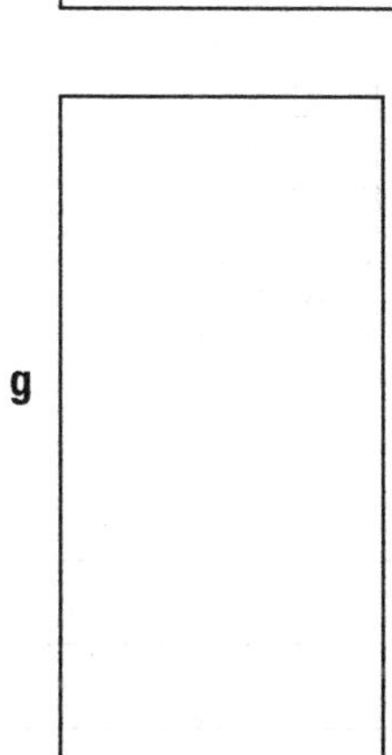

h

6.2.4 Find the area of composite shapes

6.2.5 Investigate and use area rules for triangles and rectangles

Calculate the area of rectangles

1 Use the measurements to find the area of each rectangle.

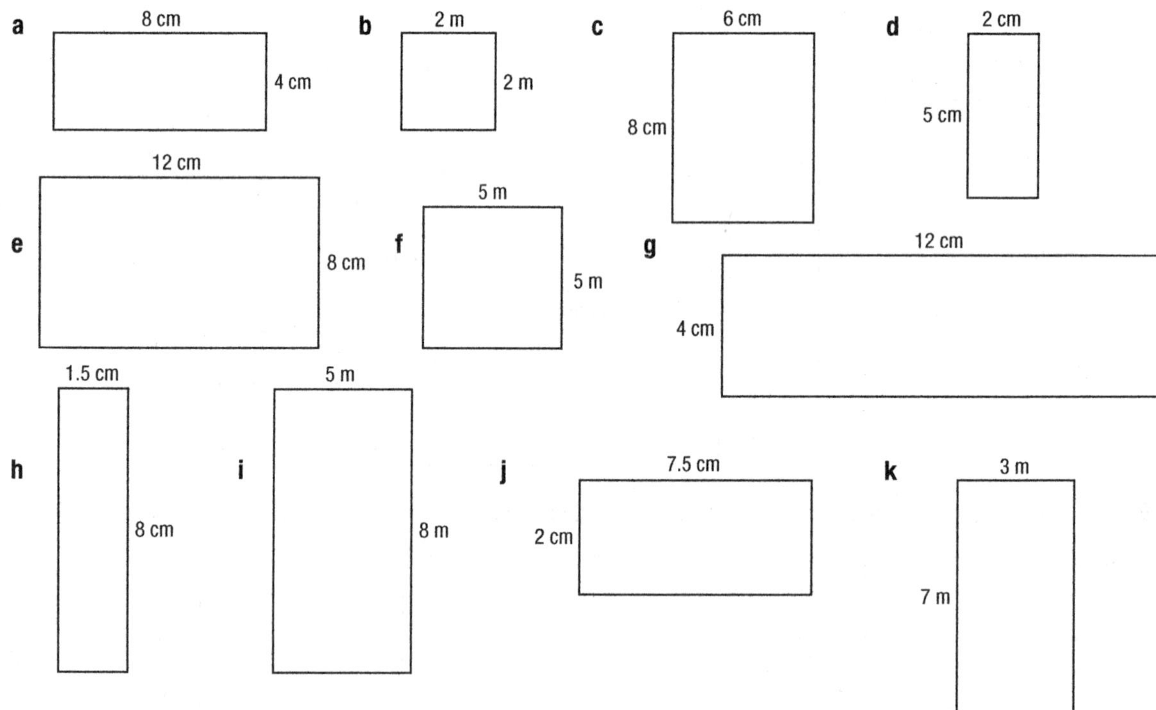

2 Copy and complete the chart.

	Length	Width	Area
a	5 cm	7 cm	
b	6 cm		36 cm^2
c		3 cm	15 cm^2
d	8 cm		48 cm^2
e	10 cm	4 cm	
f		5 cm	20 cm^2

	Length	Width	Area
g	9 cm	6 cm	
h	11 cm		55 cm^2
i		2 cm	24 cm^2
j		7 cm	63 cm^2
k	8 cm	9 cm	
l	12 cm		60 cm^2

3 Work out the dimensions in whole centimetres of all possible rectangles with an area of:

a 36 cm^2 ______________________

b 16 cm^2 ______________________

c 40 cm^2 ______________________

d 18 cm^2 ______________________

Calculate the area of triangles

Help Box

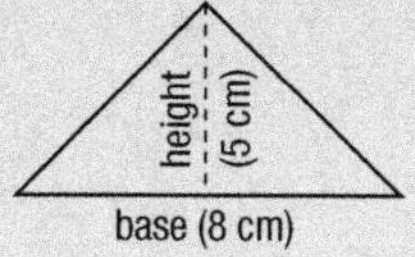

To find the area of a triangle, halve the product of its height and base.

Height × base = 5 × 8 = 40 cm

Half of 40 = 20, so the area is 20 cm^2.

Use the measurements to find the area of each triangle.

1 height 5 cm, base 6 cm

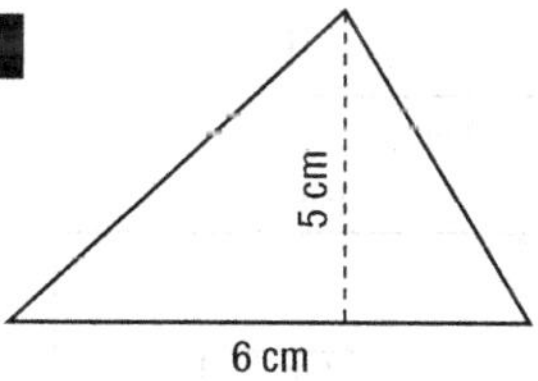

2 height 2 cm, base 8 cm

3 height 7 cm, base 5 cm

4 height 6 cm, base 7 cm

5 height 10 cm, base 5 cm

6 height 4 cm, base 11 cm

7 height 4 cm, base 8 cm

8 height 5 cm, base 8 cm

9 height 12 cm, base 10 cm

10 height 2 cm, base 13 cm

11 height 13 cm, base 5 cm

12 height 3 cm, base 7 cm

13 height 15 cm, base 12 cm

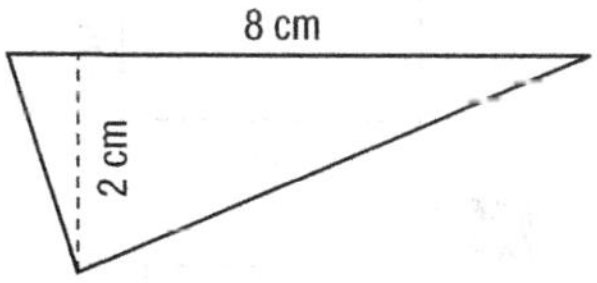

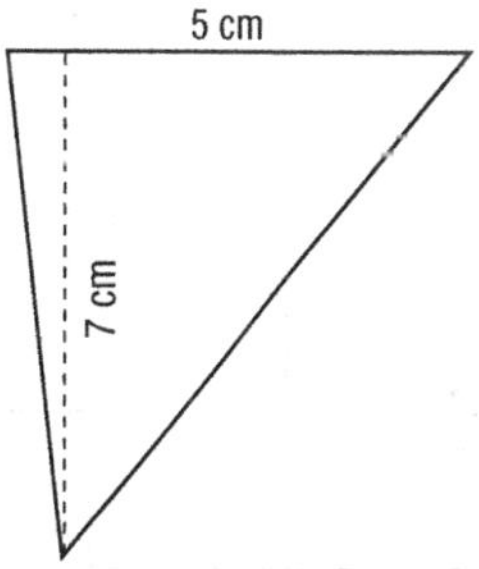

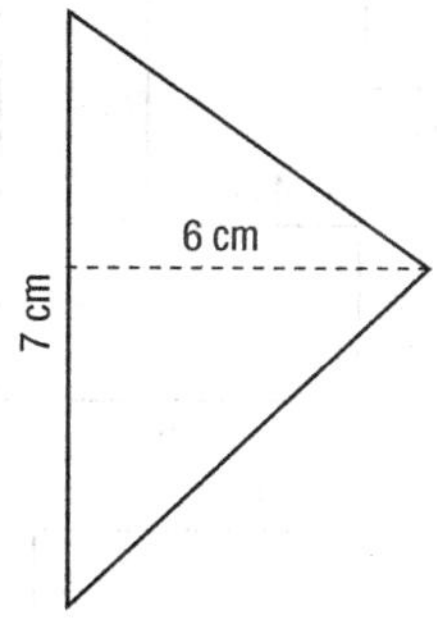

14 height 4 cm, base 8.5 cm

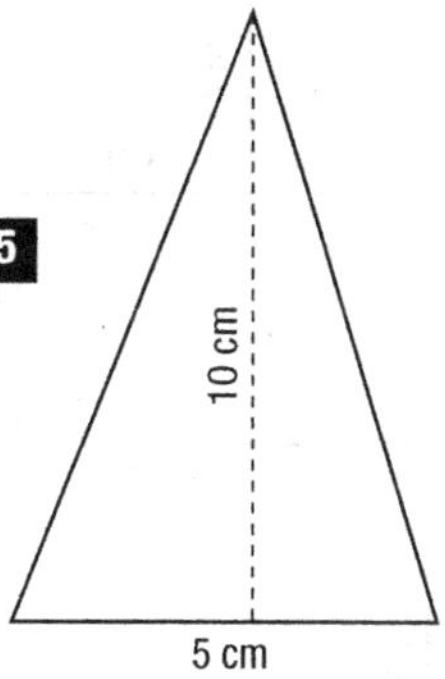

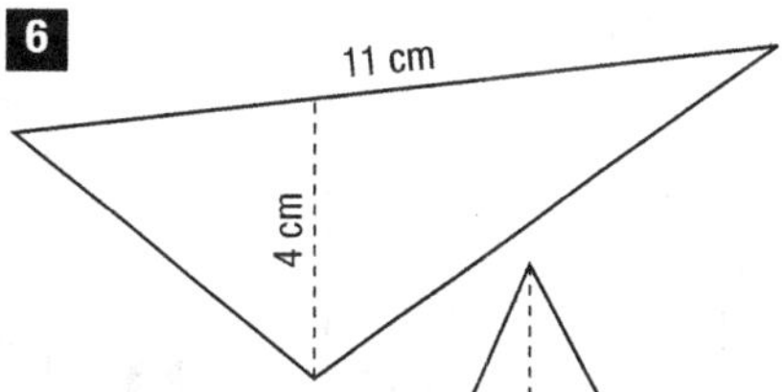

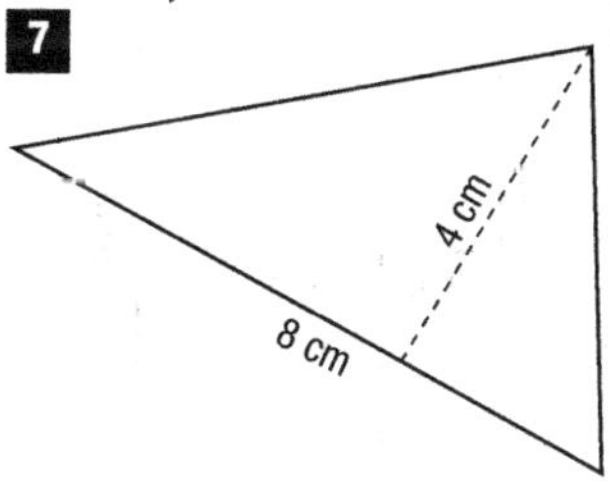

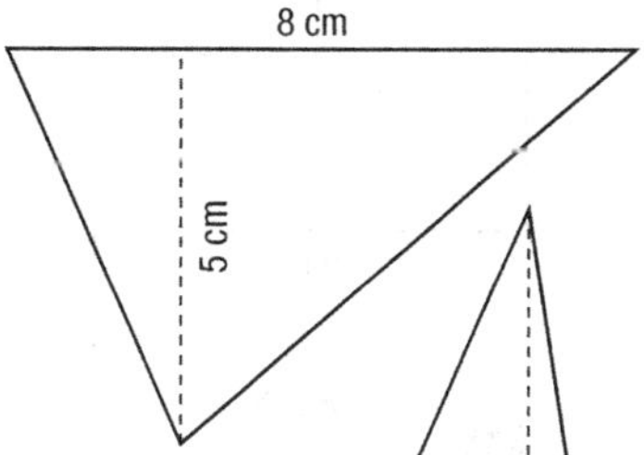

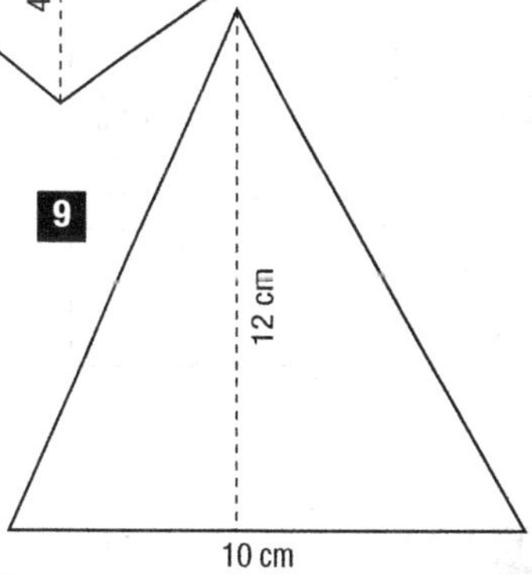

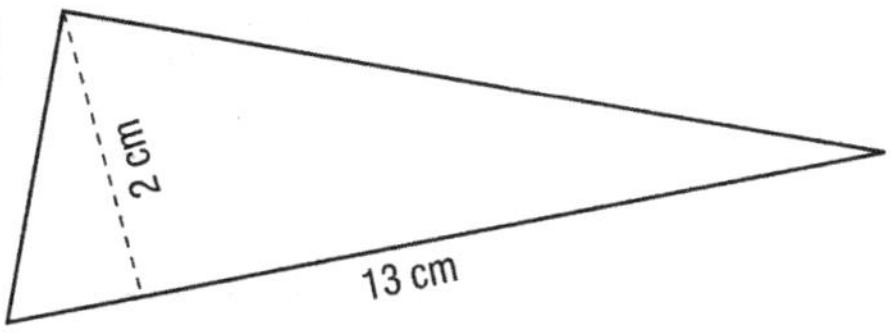

15 height 5 cm, base 15 cm

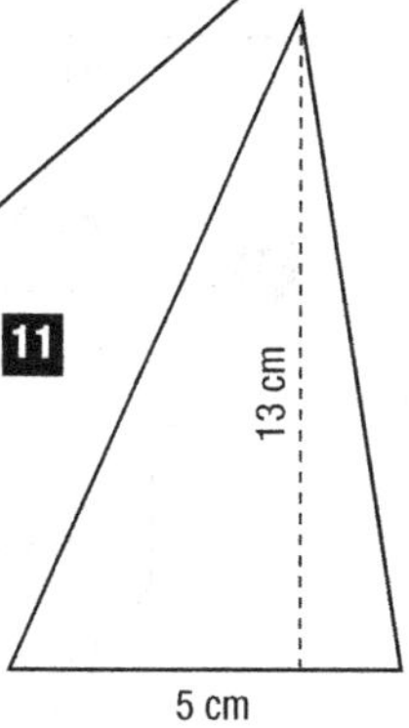

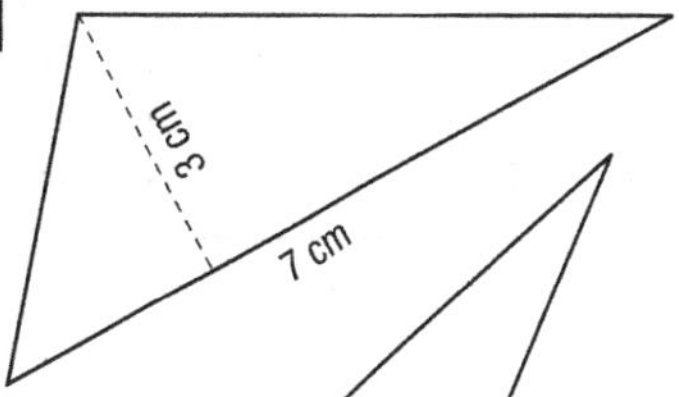

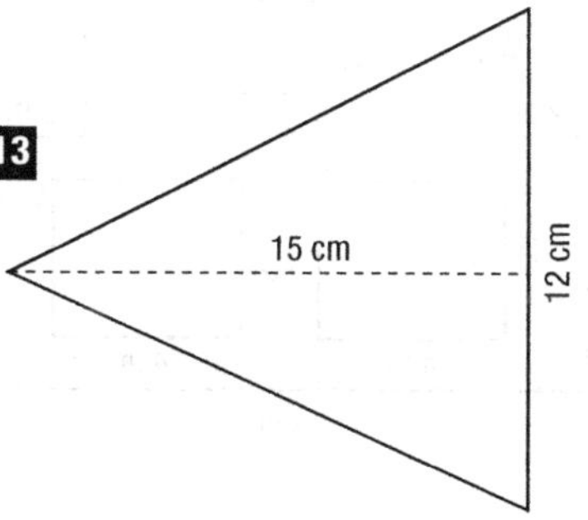

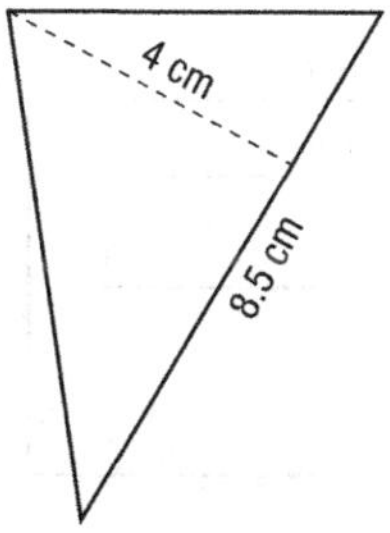

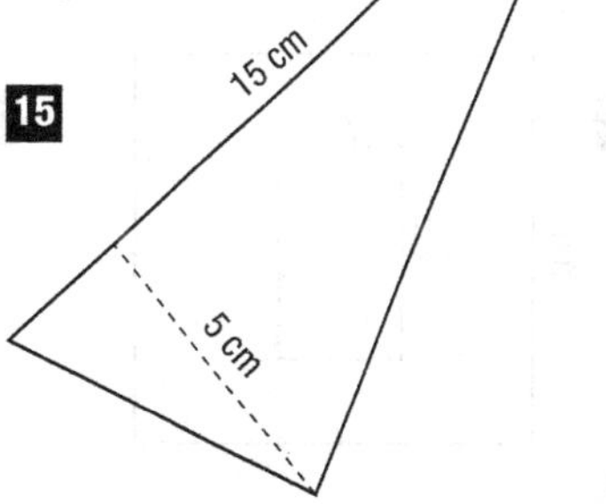

Calculate the area of composite shapes

Use the measurements to find the area of the shaded part of each shape.

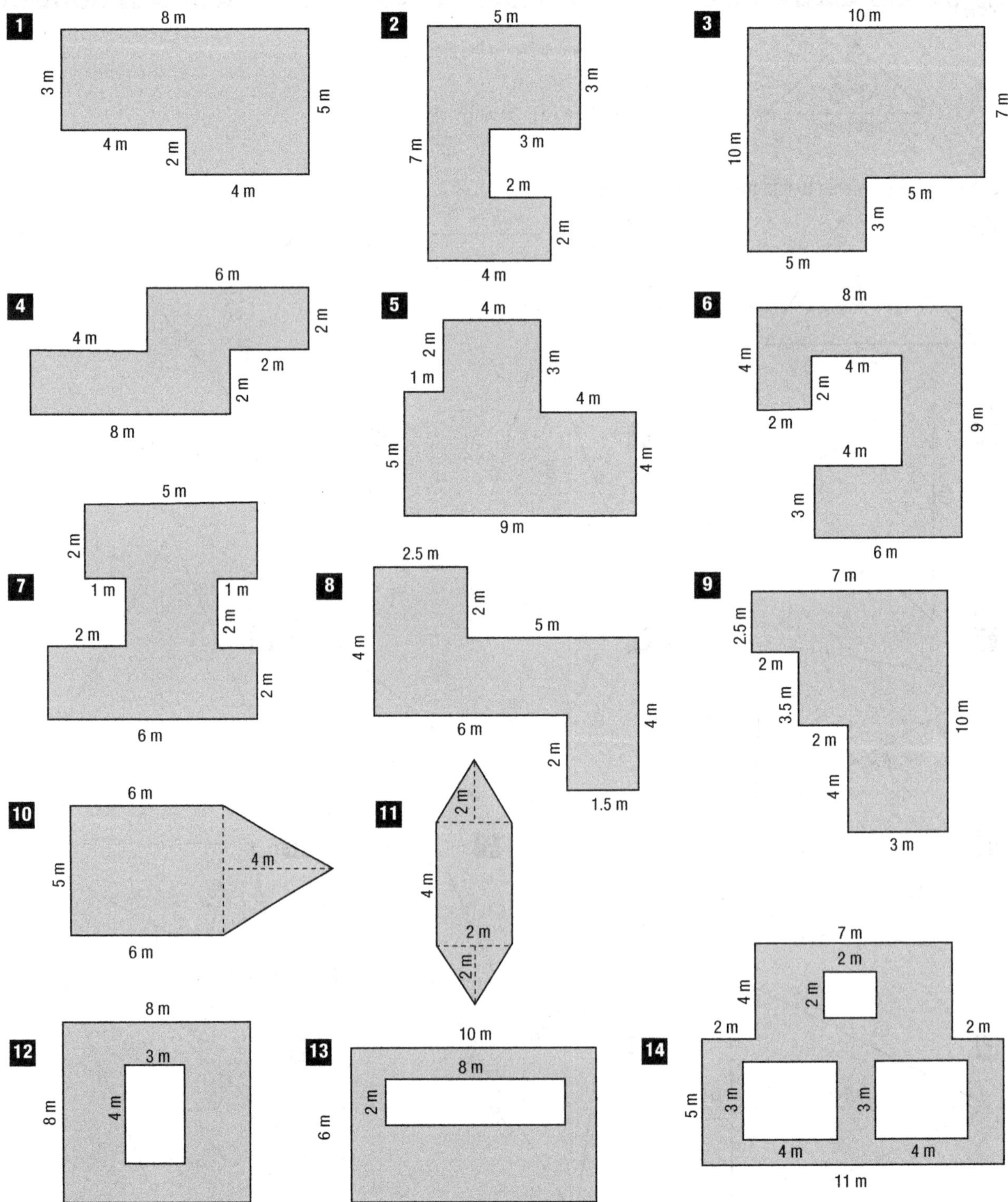

6.2.6 Investigate volumes of simple solids to determine rules

Calculate the volume of rectangular prisms

Use the measurements to find the volume of each rectangular prism.

1 5 cm, 2 cm, 2 cm

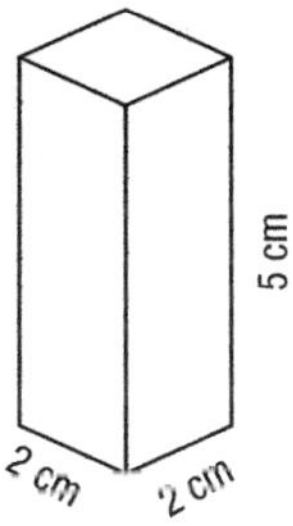

2 4 cm, 5 cm, 3 cm

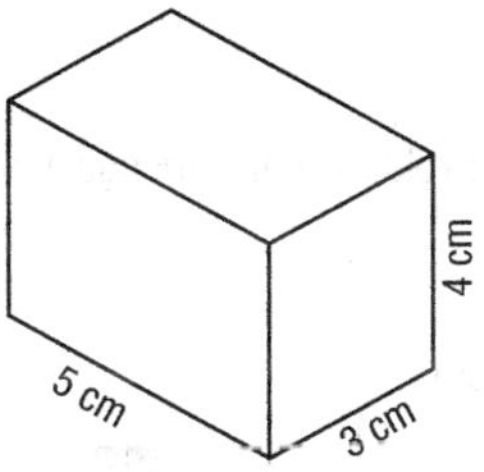

3 1 cm, 6 cm, 6 cm

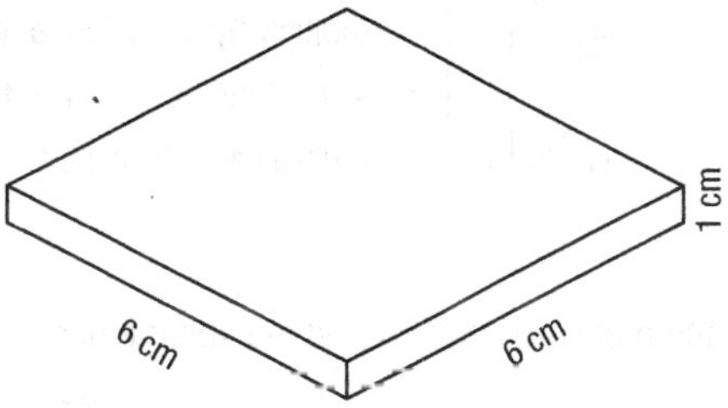

4 2 cm, 2 cm, 8 cm

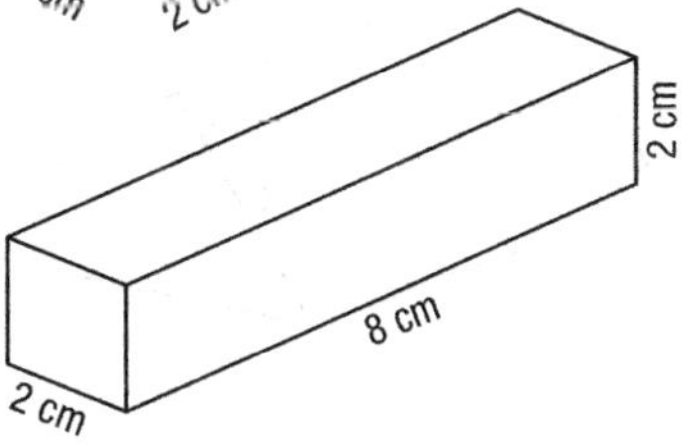

5 8 cm, 4 cm, 1 cm

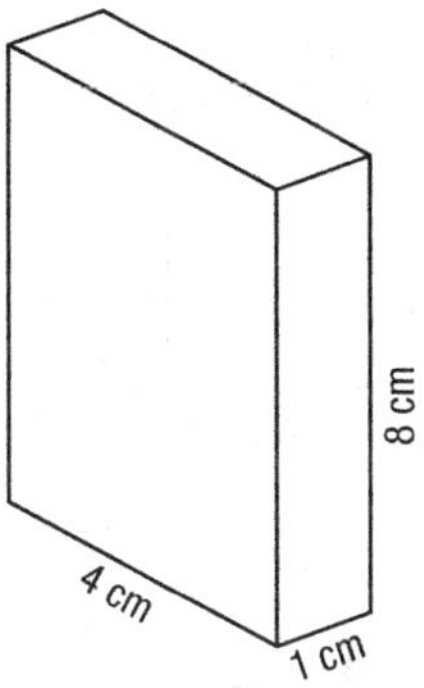

6 6 cm, 6 cm, 2 cm

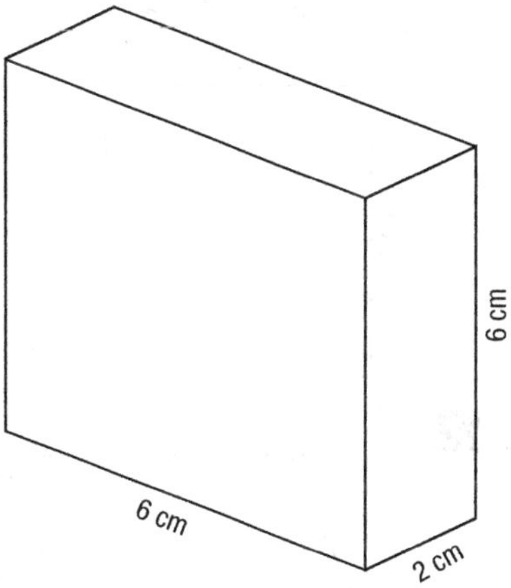

7 5 cm, 4 cm, 4 cm

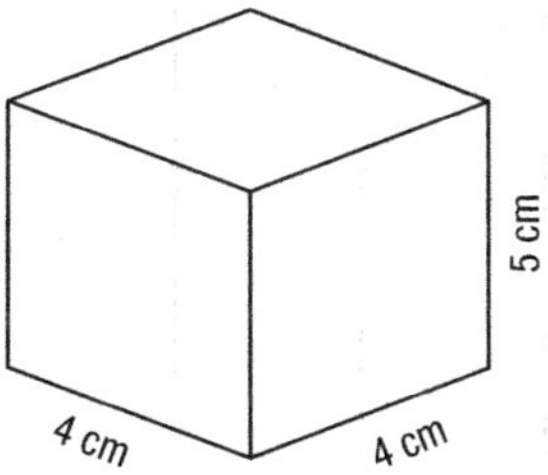

8 1 cm, 2 cm, 6 cm

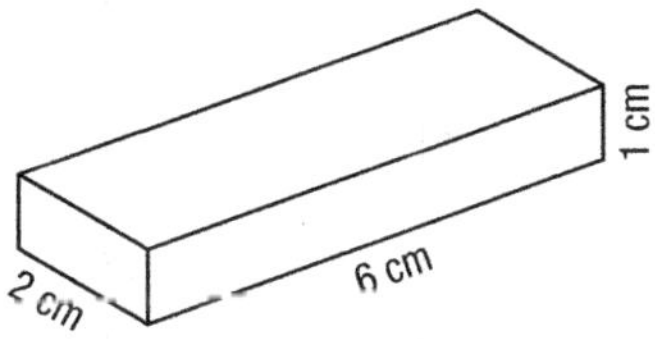

9 4 cm, 3 cm, 2 cm

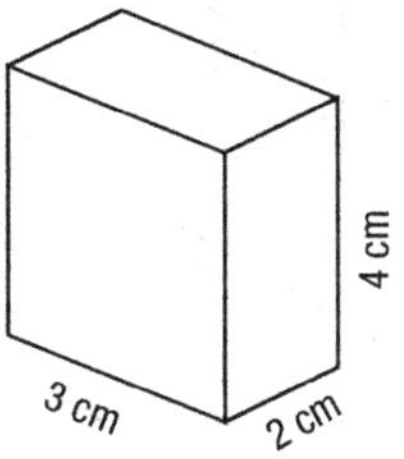

10 4 cm, 7 cm, 3 cm

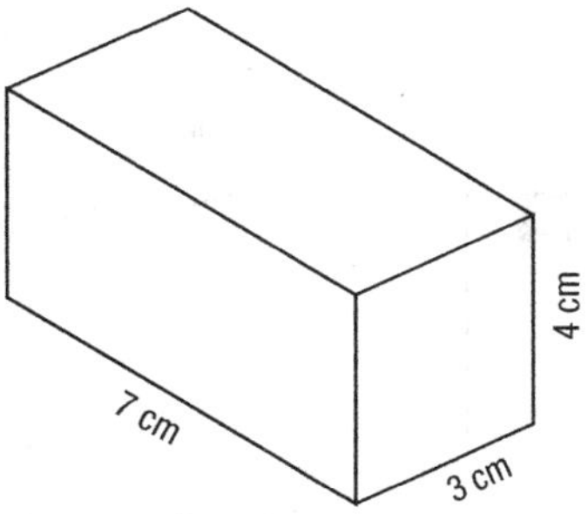

11 9 cm, 1 cm, 5 cm

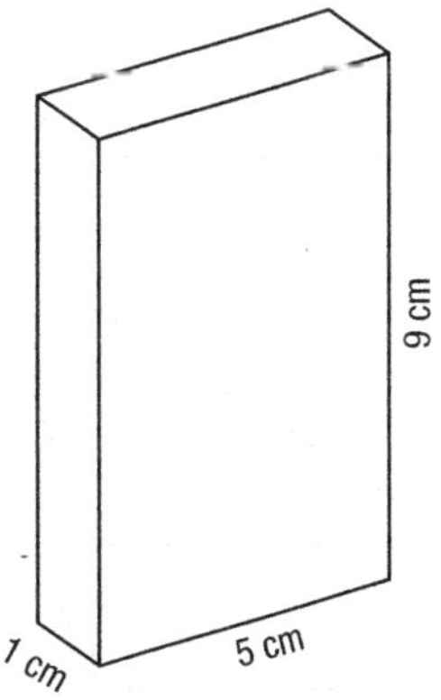

12 1.5 cm, 1 cm, 6 cm

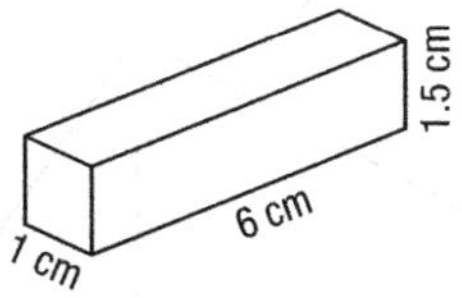

13 2 cm, 5 cm, 6 cm

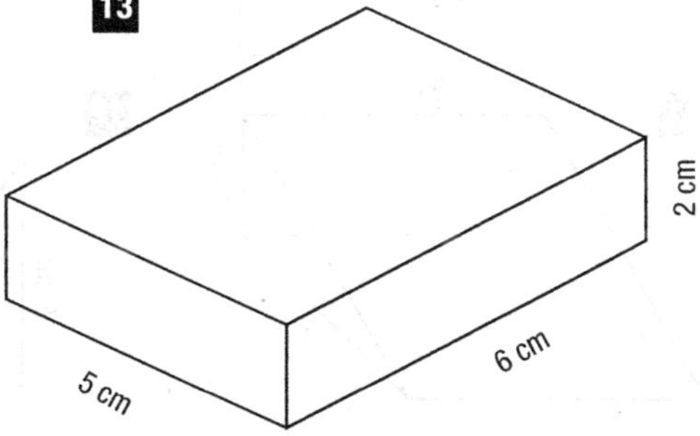

14 3 cm, 5 cm, 3 cm

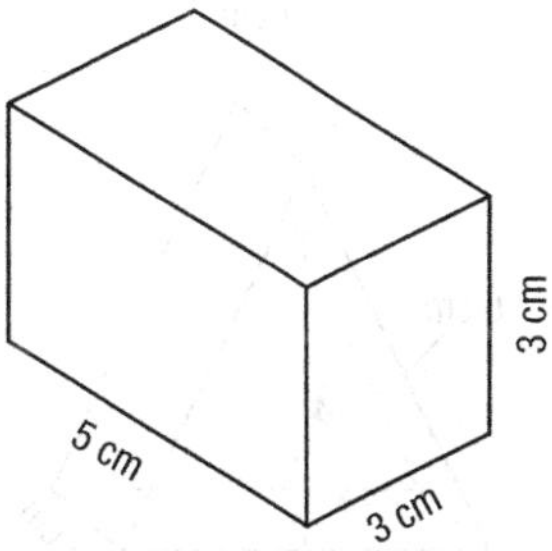

Calculate the volume of triangular prisms

Help Box

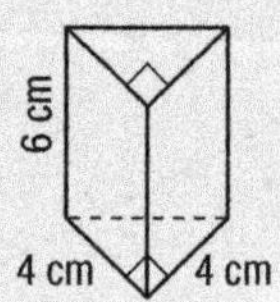

To find the volume of a triangular prism, find the area of the triangular base and then multiply by the height of the prism.

Area of the base = $(4 \times 4) \div 2 = 8$

Multiply area by height: $8 \times 6 = 48$, so the volume of this triangular prism is 48 cm^3.

Use the measurements to find the volume of each triangular prism.

1 5 cm, 5 cm, 8 cm

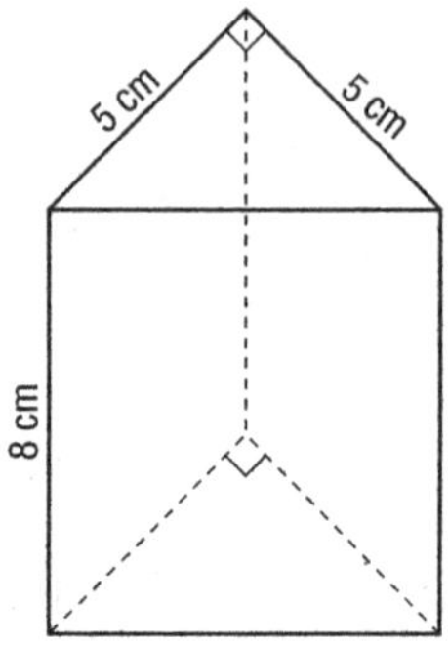

2 5 cm, 4 cm, 5 cm

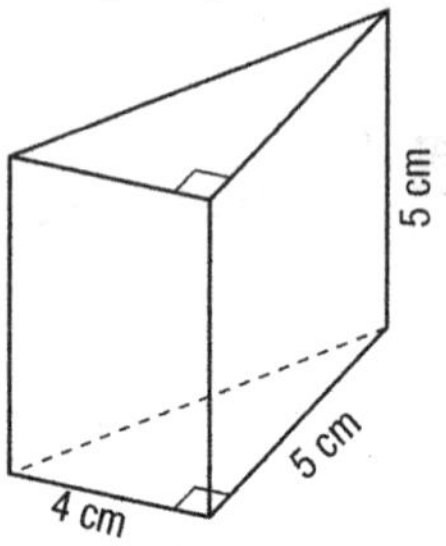

3 7 cm, 5 cm, 6 cm

4 11 cm, 6 cm, 8 cm

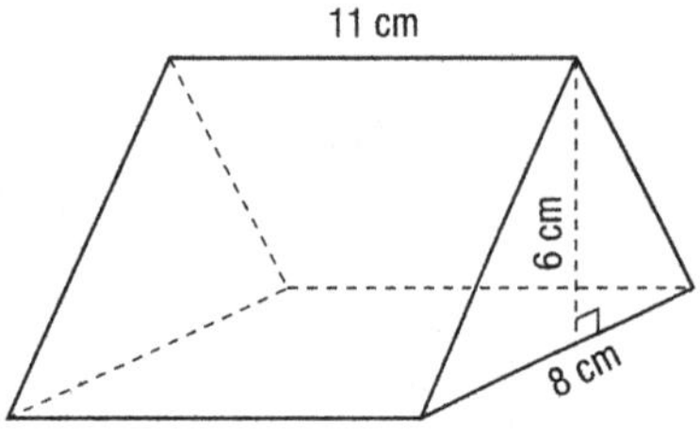

5 4 cm, 3 cm, 10 cm

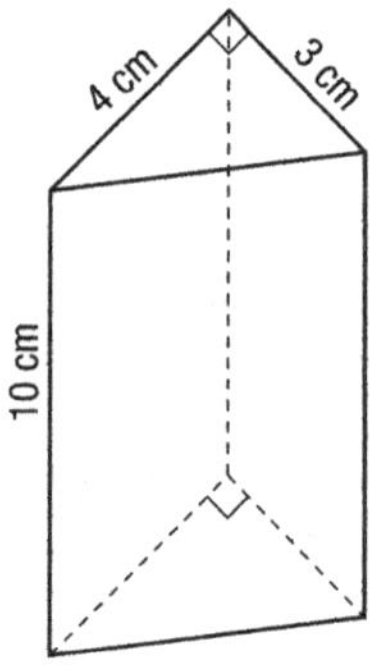

6 12 cm, 5 cm, 5 cm

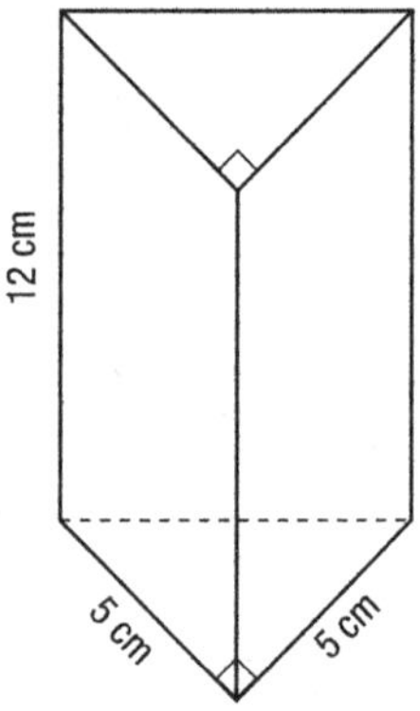

7 2 cm, 7 cm, 8 cm

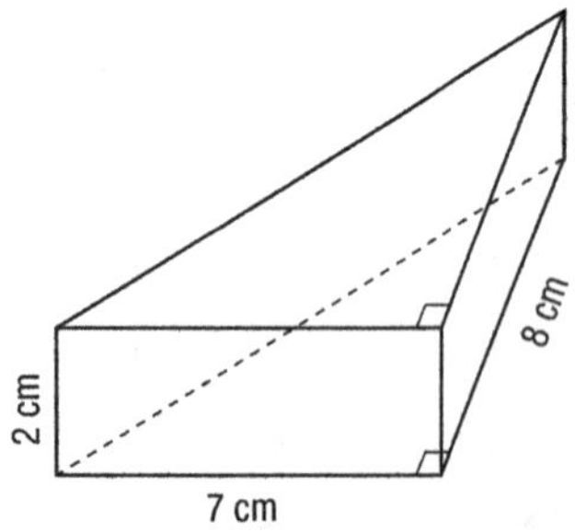

8 5 cm, 8 cm, 9 cm

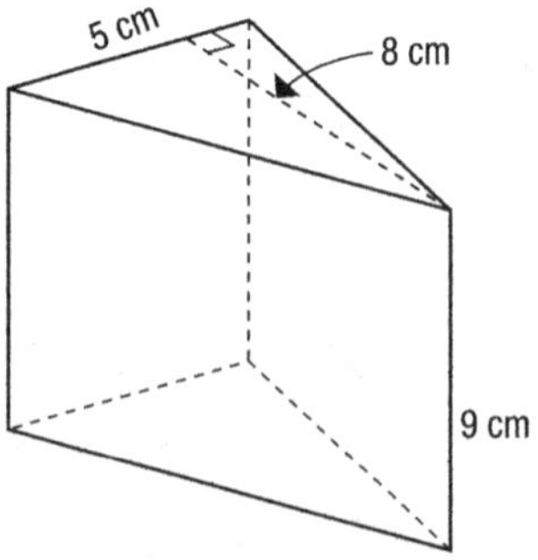

9 12 cm, 3 cm, 3 cm

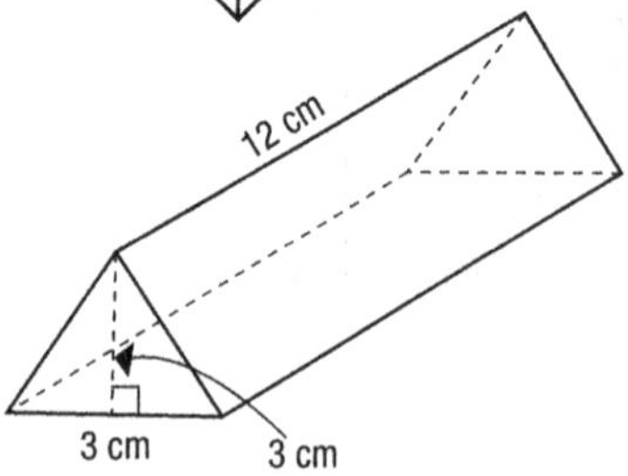

10 6 cm, 6 cm, 1.5 cm

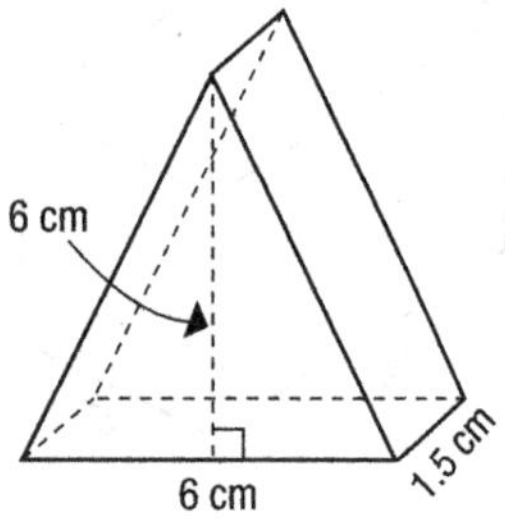

11 6 cm, 2 cm, 5 cm

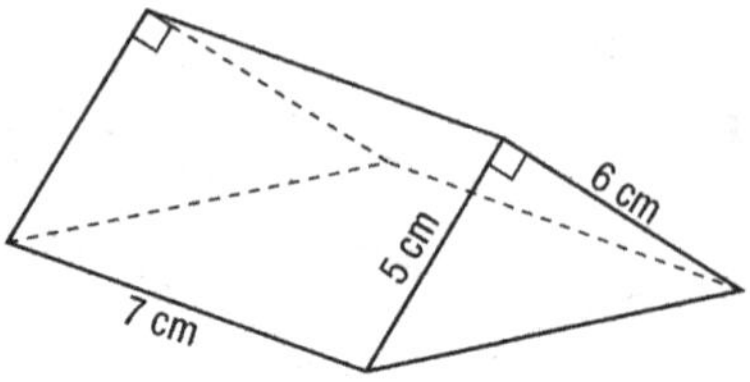

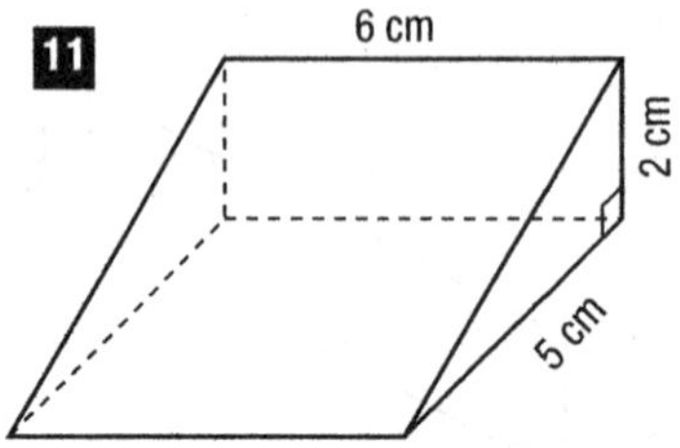

12 5 cm, 3 cm, 10 cm

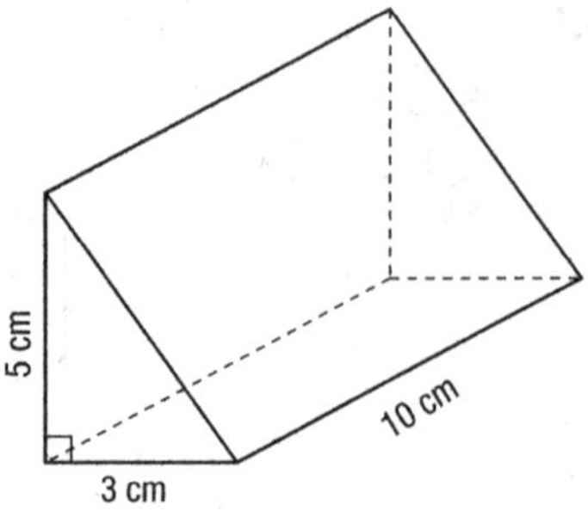

6.2.8 Convert between metric units

Convert between metric units of capacity

1 Write each quantity as millilitres, for example: 3 L 400 mL = 3400 mL

a 1 L 560 mL **b** 3 L 450 mL **c** 2 L 125 mL **d** 5 L 615 mL
e 2 L 975 mL **f** 10 L 723 mL **g** 4 L 364 mL **h** 8 L 45 mL
i 12 L 75 mL **j** 1 L 30 mL **k** 3 L 5 mL **l** 6 L 8 mL

2 Write each quantity as litres and millilitres, for example: 5725 mL = 5 L 725 mL

a 5641 mL **b** 2200 mL **c** 1759 mL **d** 3085 mL
e 1060 mL **f** 7005 mL **g** 817 mL **h** 384 mL
i 12 356 mL **j** 10 095 mL **k** 15 208 mL **l** 11 410 mL

3 Write these litre (L) quantities as millilitres (mL).

a 1.5 L **b** 2.75 L **c** 0.2 L **d** 1.25 L **e** 3.5 L **f** 5.625 L
g 0.815 L **h** 3.27 L **i** 1.6 L **j** 7.36 L **k** 9.7 L **l** 6.05 L
m 3.619 L **n** 2.8 L **o** 5.5 L **p** 0.429 L **q** 1.987 L **r** 0.25 L
s 2.311 L **t** 0.08 L **u** 1.34 L **v** 7.06 L **w** 3.088 L **x** 4.39 L

4 Use decimal notation to write these quantities as litres (L).

a 2500 mL **b** 4500 mL **c** 1750 mL **d** 465 mL **e** 275 mL
f 1075 mL **g** 218 mL **h** 32 mL **i** 1800 mL **j** 1030 mL
k 410 mL **l** 5400 mL **m** 875 mL **n** 2300 mL **o** 4010 mL
p 75 mL **q** 40 mL **r** 5109 mL **s** 633 mL **t** 3002 mL

5 How many millilitres?

a $5\frac{1}{2}$ L **b** $\frac{1}{4}$ L **c** $1\frac{1}{4}$ L **d** $\frac{1}{2}$ L **e** $\frac{3}{4}$ L **f** $3\frac{3}{4}$ L
g $\frac{5}{10}$ L **h** $\frac{9}{10}$ L **i** $4\frac{1}{5}$ L **j** $2\frac{1}{4}$ L **k** $1\frac{3}{10}$ L **l** $2\frac{3}{4}$ L
m $7\frac{1}{4}$ L **n** $\frac{1}{5}$ L **o** $4\frac{4}{5}$ L **p** $\frac{1}{8}$ L **q** $3\frac{1}{8}$ L **r** $\frac{2}{10}$ L

6 Write each quantity as a fraction of a litre, for example: 7500 mL = $7\frac{1}{2}$ L

a 2500 mL **b** 3250 mL **c** 750 mL **d** 4200 mL **e** 900 mL
f 5100 mL **g** 6500 mL **h** 1750 mL **i** 3500 mL **j** 250 mL
k 125 mL **l** 5125 mL **m** 7750 mL **n** 2600 mL **o** 500 mL
p 100 mL **q** 1250 mL **r** 6125 mL **s** 8100 mL **t** 625 mL

Convert between metric units of area

1 How many square centimetres (cm^2) in these square metres (m^2)?

a	6 m^2	b	2 m^2	c	7 m^2	d	3 m^2	e	1.5 m^2
f	9.5 m^2	g	5.5 m^2	h	0.5 m^2	i	10 m^2	j	8.25 m^2
k	4.25 m^2	l	1.25 m^2	m	5.75 m^2	n	2.75 m^2	o	0.75 m^2
p	10.75 m^2	q	3.8 m^2	r	7.2 m^2	s	1.9 m^2	t	6.125 m^2
u	0.375 m^2	v	2.625 m^2	w	1.075 m^2	x	9.025 m^2		

2 Use decimal notation where necessary to write these as square metres (m^2).

a	20 000 cm^2	b	50 000 cm^2	c	170 000 cm^2	d	240 000 cm^2	e	25 000 cm^2
f	75 000 cm^2	g	45 000 cm^2	h	15 000 cm^2	i	30 500 cm^2	j	18 500 cm^2
k	15 230 cm^2	l	27 180 cm^2	m	8500 cm^2	n	5890 cm^2	o	43 054 cm^2
p	75 140 cm^2	q	102 650 cm^2	r	340 080 cm^2	s	3005 cm^2	t	9670 cm^2

3 How many square metres (m^2) in these hectares (ha)?

a	4 ha	b	7 ha	c	15 ha	d	11 ha	e	8.5 ha
f	3.5 ha	g	10.5 ha	h	1.5 ha	i	3.25 ha	j	12.25 ha
k	5.75 ha	l	9.75 ha	m	6.2 ha	n	4.8 ha	o	11.7 ha
p	2.9 ha	q	0.3 ha	r	8.125 ha	s	7.625 ha	t	0.067 ha
u	2.97 ha	v	16.35 ha	w	1.008 ha	x	3.504 ha		

4 Use decimal notation where necessary to write these as hectares (ha).

a	20 000 m^2	b	80 000 m^2	c	130 000 m^2	d	15 000 m^2	e	65 000 m^2
f	235 000 m^2	g	62 500 m^2	h	37 500 m^2	i	83 500 m^2	j	49 250 m^2
k	51 750 m^2	l	70 750 m^2	m	48 650 m^2	n	27 050 m^2	o	126 280 m^2
p	100 700 m^2	q	210 600 m^2	r	88 900 m^2	s	9000 m^2	t	7250 m^2

5 Complete the gaps.

a	$3\frac{1}{4}$ ha = ____ m^2	b	$2\frac{3}{4}$ m^2 = ____ cm^2	c	$7\frac{3}{5}$ ha = ____ m^2
d	$5\frac{1}{4}$ m^2 = ____ cm^2	e	$\frac{1}{2}$ ha = ____ m^2	f	$6\frac{1}{2}$ m^2 = ____ cm^2
g	$\frac{7}{10}$ ha = ____ m^2	h	$4\frac{2}{5}$ m^2 = ____ cm^2	i	$10\frac{1}{4}$ ha = ____ m^2
j	$\frac{4}{5}$ m^2 = ____ cm^2	k	$1\frac{9}{10}$ ha = ____ m^2	l	$2\frac{1}{8}$ m^2 = ____ cm^2
m	$5\frac{3}{8}$ ha = ____ m^2	n	$7\frac{1}{8}$ m^2 = ____ cm^2	o	$4\frac{5}{8}$ ha = ____ m^2
p	$\frac{7}{8}$ m^2 = ____ cm^2	q	$3\frac{1}{20}$ ha = ____ m^2	r	$8\frac{1}{50}$ m^2 = ____ cm^2
s	$11\frac{13}{20}$ ha = ____ m^2	t	$6\frac{17}{25}$ m^2 = ____ cm^2	u	$5\frac{9}{20}$ ha = ____ m^2

6.2.11 Identify different angles

Identify different angles

1 Use **acute**, **obtuse**, **right**, **straight** or **reflex** to name each angle.

a
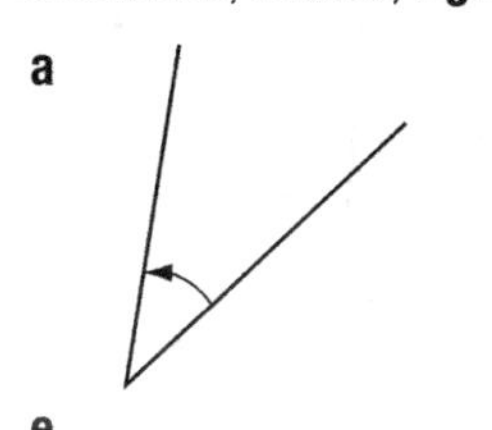

b

c
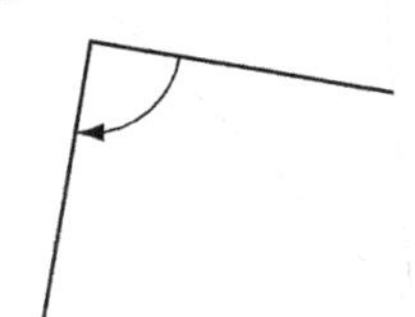

d

e
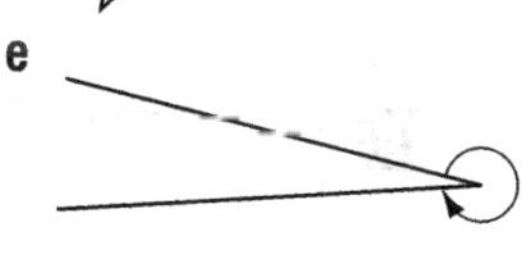

f

g
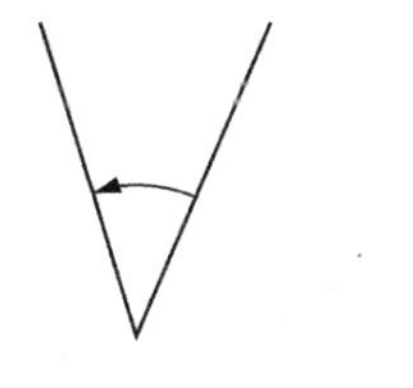

h
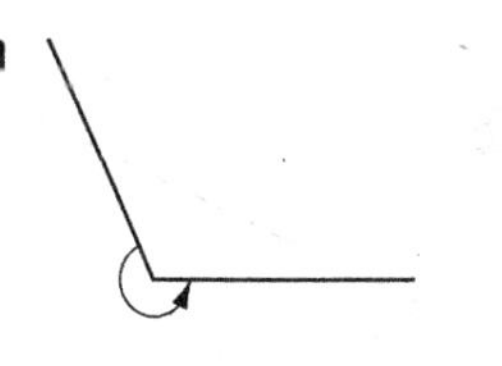

i
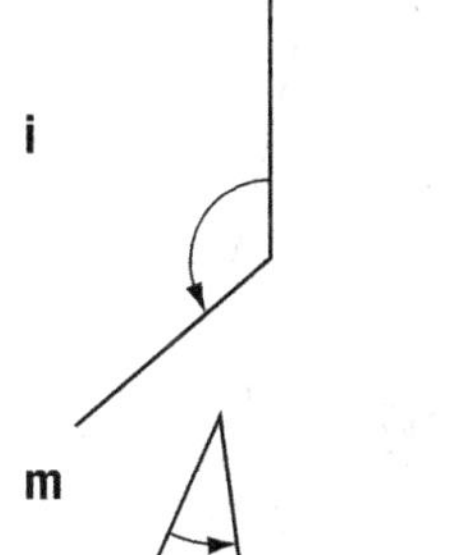

j

k

l

m

n
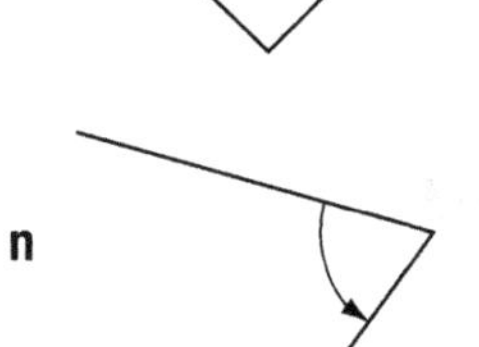

o
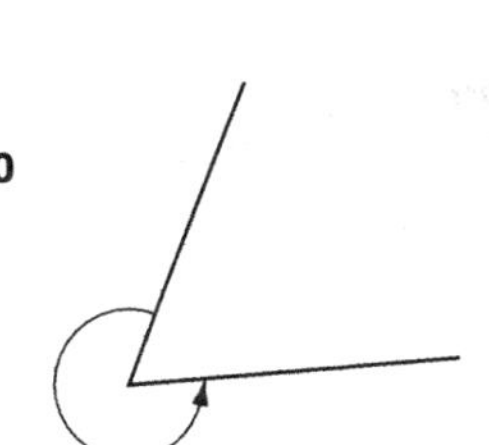

p
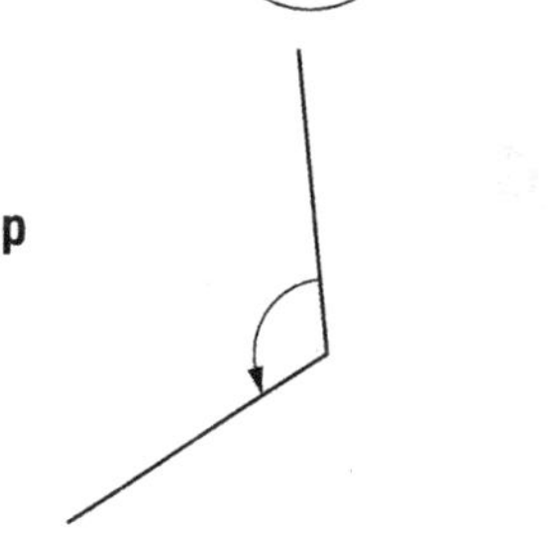

2 Name the angle that is marked in each shape.

a
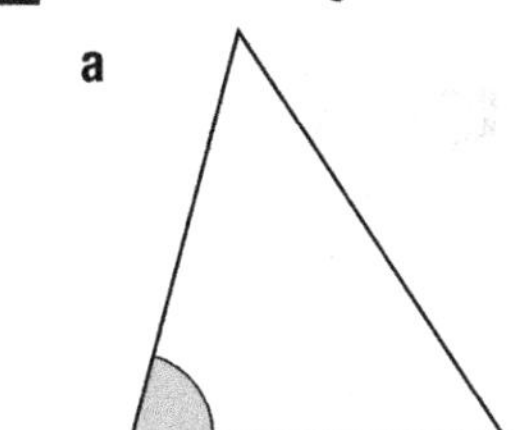

b
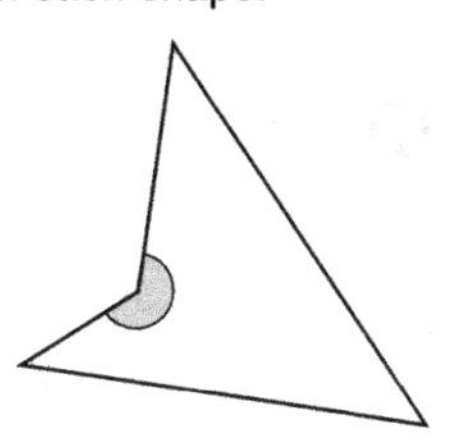

c
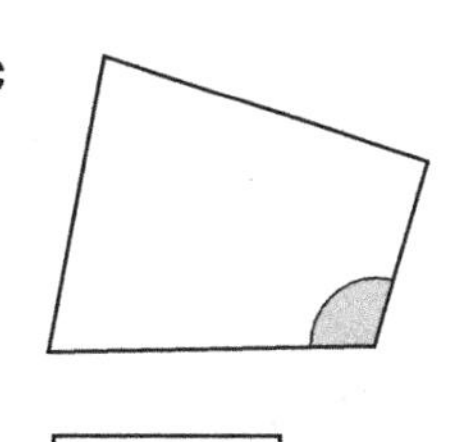

d
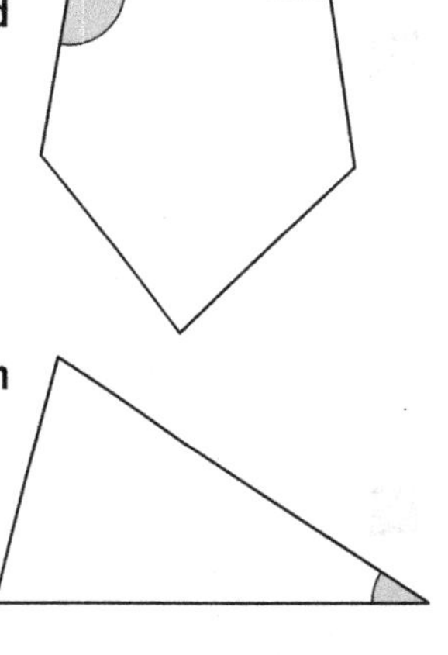

e
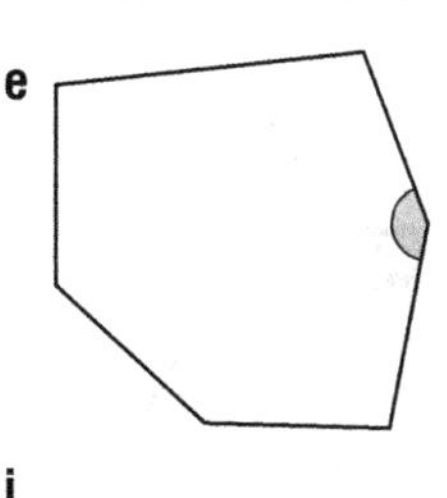

f
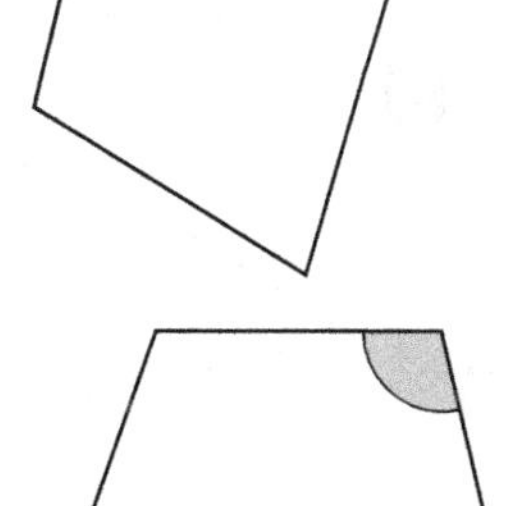

g

h

i

j

k
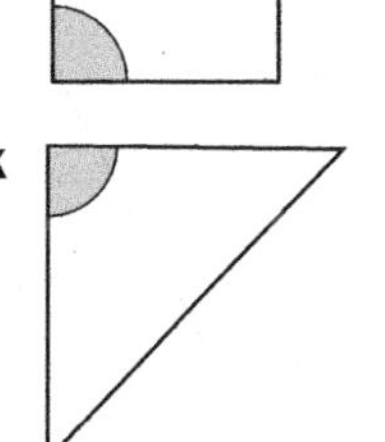

l

Measure angles

Use a protractor to measure each of the angles below.

1

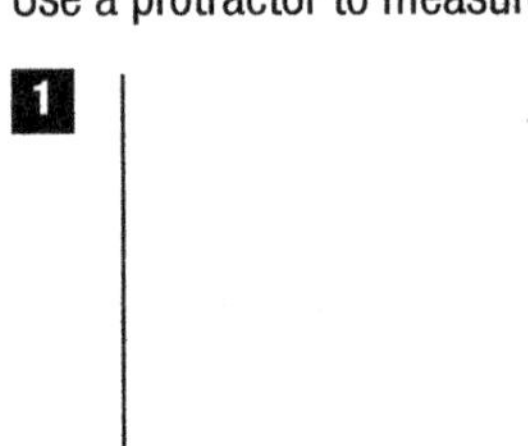

2

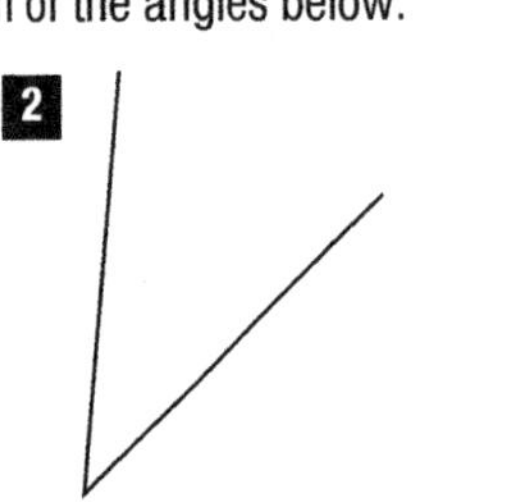

3

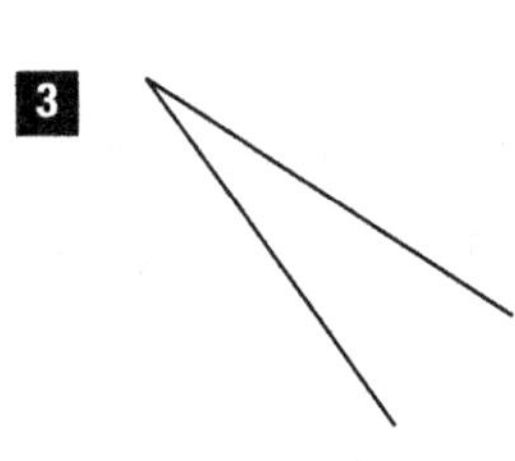

4

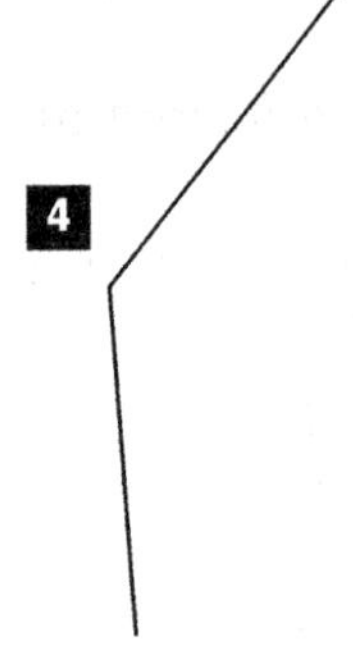

5

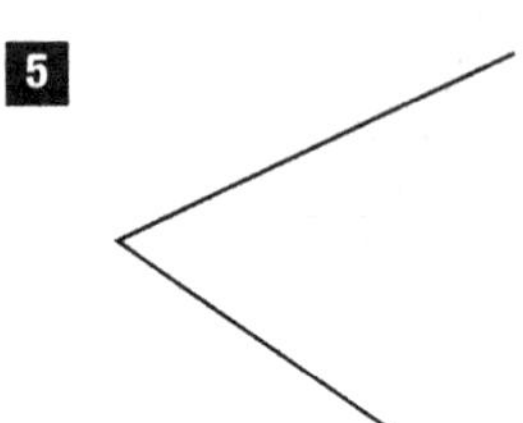

6

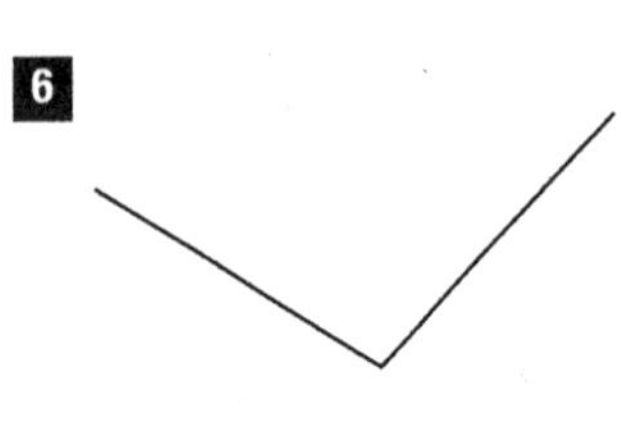

7

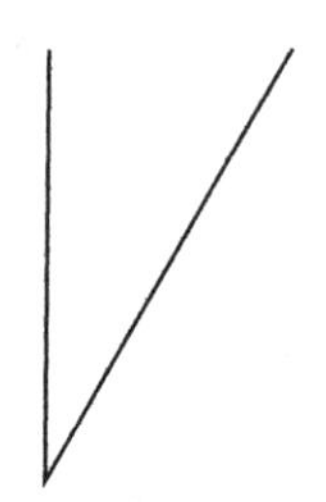

8

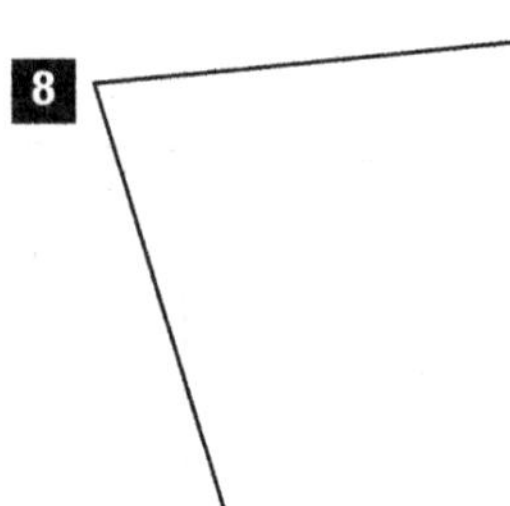

9

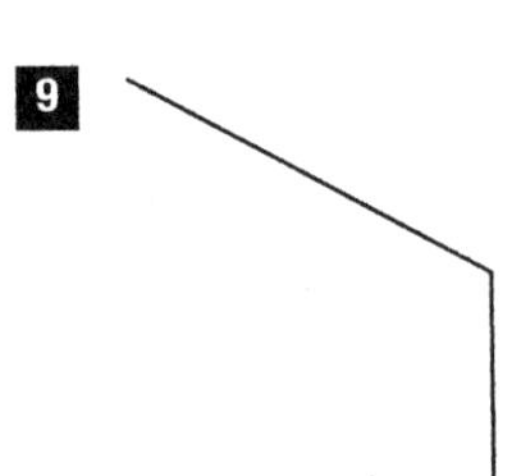

10

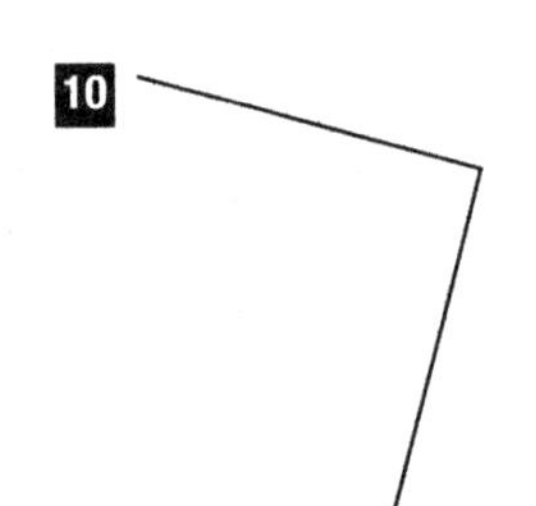

11

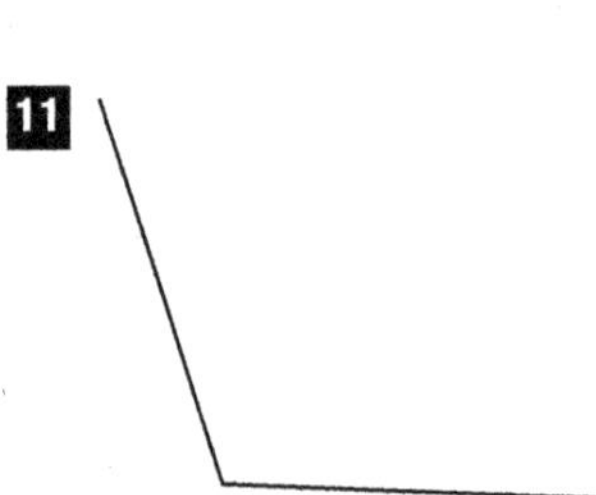

12

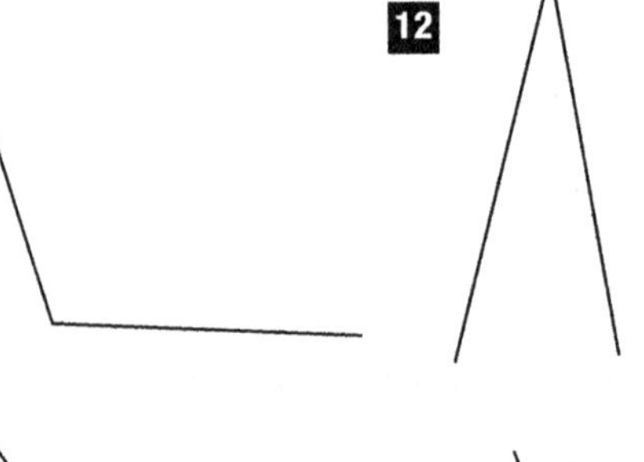

13

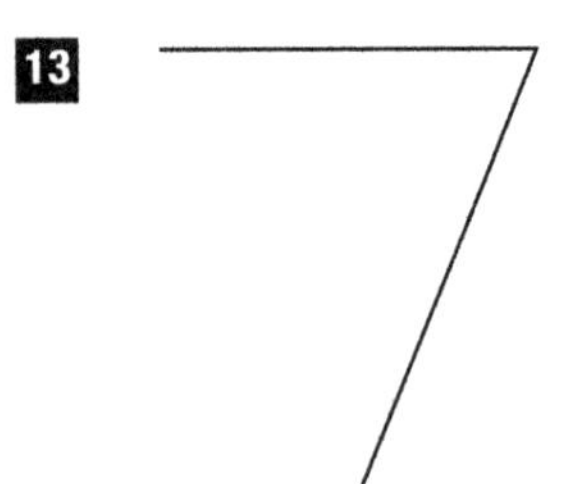

14

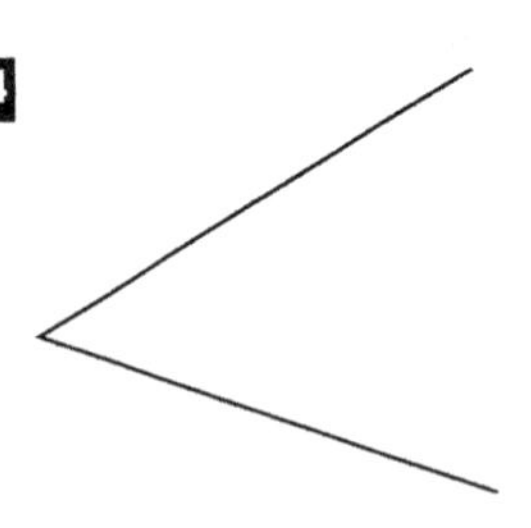

15

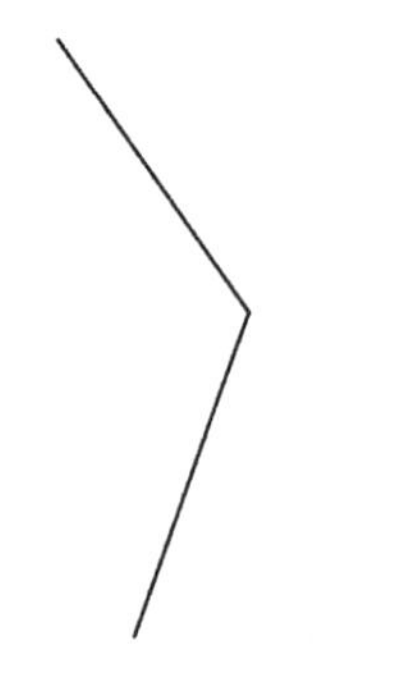

16

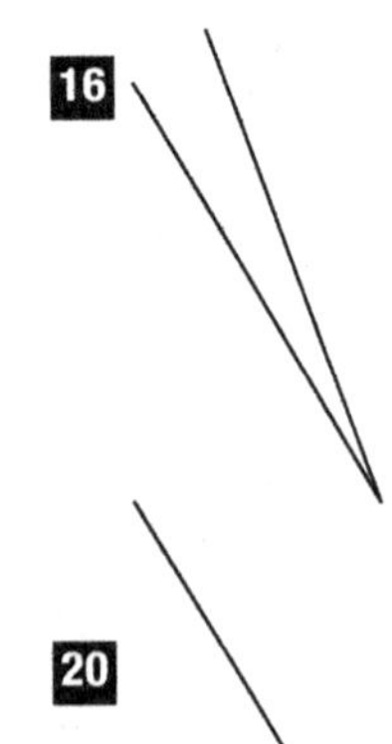

17

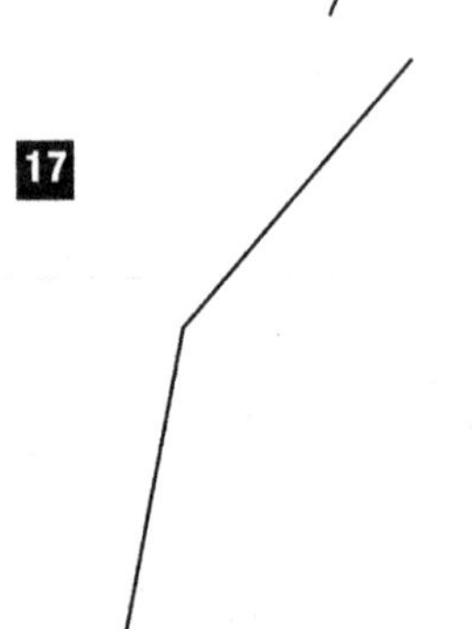

18

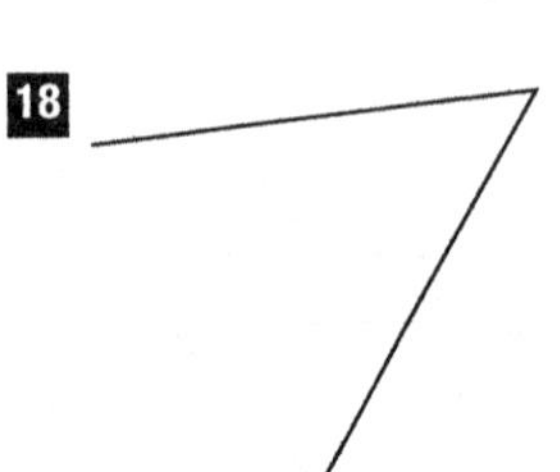

19

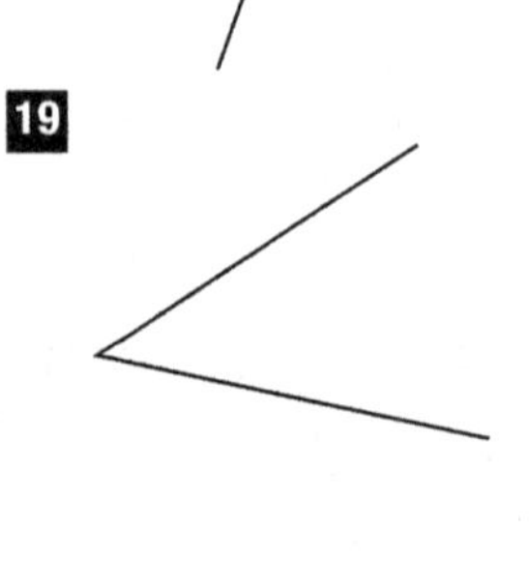

20

Find the size of unknown angles

Calculate the missing angles.

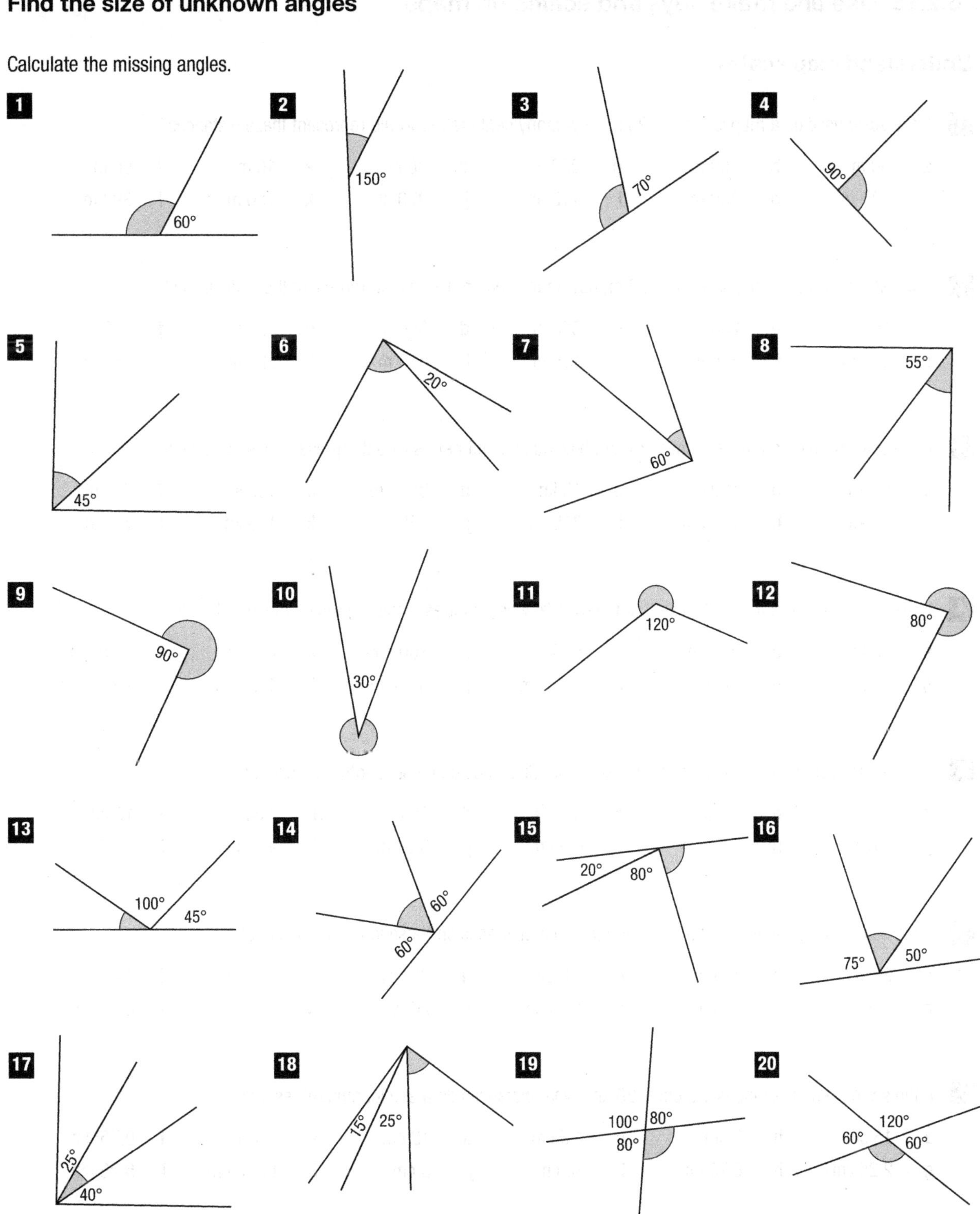

6.2.15 Use and make keys and scales on maps

Understand map scales

1 If the scale used on a map is 1 cm = 20 m, how many centimetres would represent these distances?

a	60 m	**b**	100 m	**c**	200 m	**d**	30 m	**e**	10 m	**f**	50 m
g	130 m	**h**	500 m	**i**	110 m	**j**	270 m	**k**	150 m	**l**	390 m

2 If the scale used on a map is 1 cm = 50 m, how many centimetres would represent these distances?

a	100 m	**b**	200 m	**c**	500 m	**d**	350 m	**e**	25 m	**f**	150 m
g	225 m	**h**	375 m	**i**	125 m	**j**	675 m	**k**	825 m	**l**	450 m

3 If the scale used on a map is 1 cm = 30 km, how many centimetres would represent these distances?

a	60 km	**b**	120 km	**c**	15 km	**d**	300 km	**e**	150 km	**f**	75 km
g	45 km	**h**	135 km	**i**	315 km	**j**	255 km	**k**	105 km	**l**	375 km

4 If the scale used on a map is 1 cm = 5 km, how many centimetres would represent these distances?

a	15 km	**b**	2.5 km	**c**	50 km	**d**	100 km	**e**	7.5 km	**f**	12.5 km
g	35 km	**h**	52.5 km	**i**	22.5 km	**j**	37.5 km	**k**	42.5 km	**l**	107.5 km

5 If the scale used on a map is 2 cm = 100 km, what distances do these lengths represent?

a	4 cm	**b**	1 cm	**c**	0.5 cm	**d**	10 cm	**e**	7 cm	**f**	15 cm
g	8 cm	**h**	2.5 cm	**i**	6.5 cm	**j**	3.5 cm	**k**	9.5 cm	**l**	11.5 cm

6 If the scale used on a map is 2.5 cm = 50 km, what distances do these lengths represent?

a	5 cm	**b**	10 cm	**c**	7.5 cm	**d**	25 cm	**e**	1.25 cm	**f**	1 cm
g	0.5 cm	**h**	15 cm	**i**	12.5 cm	**j**	20 cm	**k**	17.5 cm	**l**	0.75 cm

7 If the scale used on a map is 1.5 cm = 20 km, what distances do these lengths represent?

a	15 cm	**b**	3 cm	**c**	4.5 cm	**d**	12 cm	**e**	7.5 cm	**f**	0.75 cm
g	2.25 cm	**h**	6.75 cm	**i**	30 cm	**j**	9 cm	**k**	11.25 cm	**l**	5.25 cm

6.2.16 Use map grids and coordinates to locate points

Use coordinates to locate points

Play this game with a friend. Each player needs six markers and a copy of the following grid. Both players place their six markers on different intersections of their grid, without the other player seeing.

Take turns to say coordinates, for example: (6, 3). If your friend has a marker at that point, they must remove it.
The first player to lose all six markers loses the game. The other player is the winner.

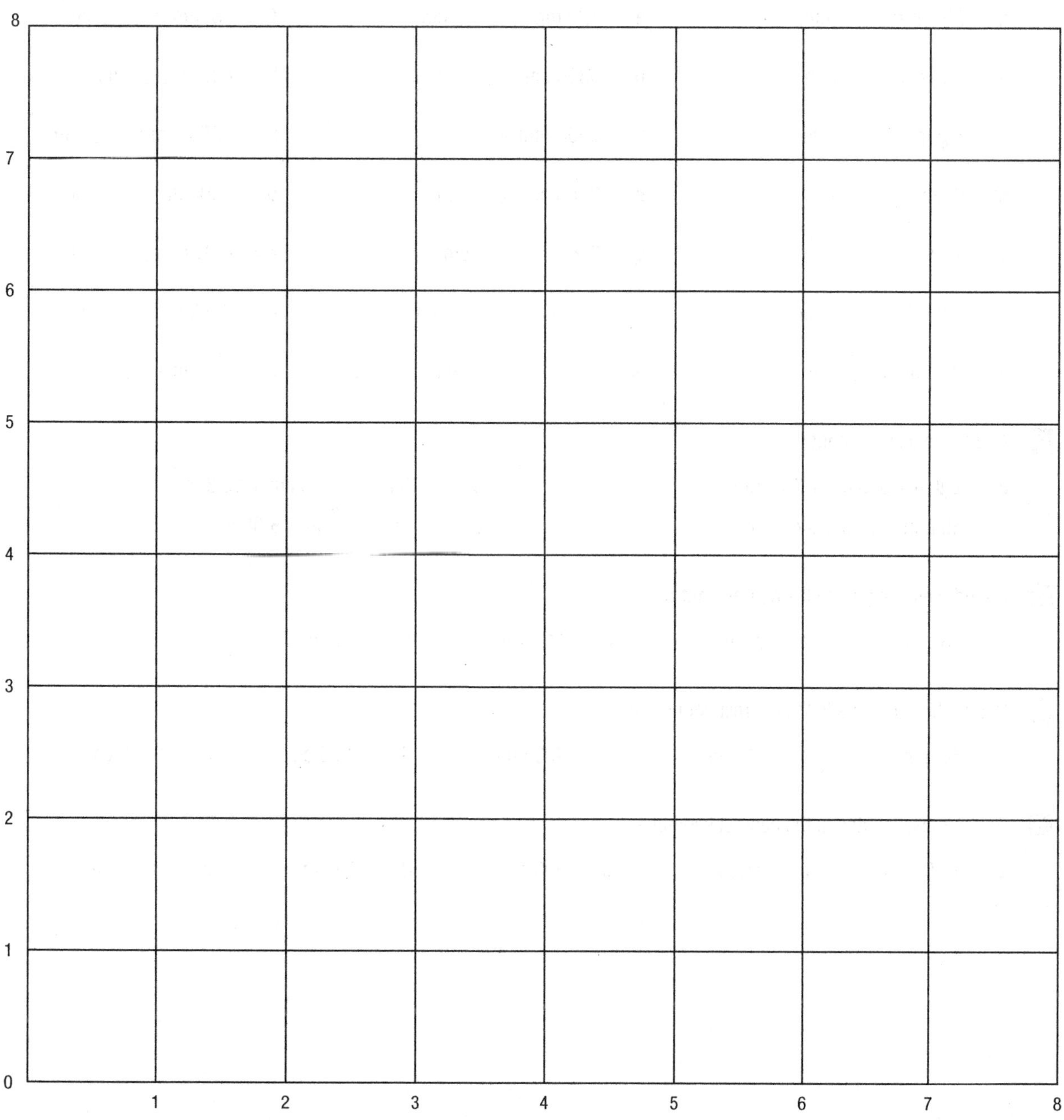

Assessment Space and Shape

1 Complete the gaps with the correct length unit.

a $4\frac{1}{2}$ m = ____ cm
b $5\frac{2}{5}$ m = ____ cm
c $7\frac{1}{2}$ cm = ____ mm
d 26 cm = ____ mm
e 420 mm = ____ cm
f 85 mm = ____ cm
g 900 cm = ____ m
h 5300 cm = ____ m
i 6 m = ____ mm
j $4\frac{3}{4}$ m = ____ mm
k 8500 mm = ____ m
l 3250 mm = ____ m
m 9 km = ____ m
n $2\frac{7}{10}$ km = ____ m
o 5500 m = ____ km
p 750 m = ____ km
q 9.2 m = ____ cm
r 6.02 m = ____ cm
s 7.8 cm = ____ mm
t 11.6 cm = ____ mm
u 8.5 km = ____ m
v 0.6 km = ____ m
w 2.7 m = ____ mm
x 5.03 m = ____ mm

2 Add these units of length.

a 2.8 m + 500 cm + 200 mm
b 1.75 km + 1500 cm + 10.3 m
c 60 mm + $\frac{7}{10}$ cm + 4.9 cm
d 800 mm + $\frac{2}{5}$ m + 5.08 m

3 Round these lengths to the nearest metre.

a 768 cm b 2726 cm c 11.04 m d 14.5 m e 1580 cm

4 Round these lengths to the nearest centimetre.

a 58 mm b 316 mm c 4035 mm d 24.85 cm e 16.09 cm

5 Round these lengths to the nearest kilometre.

a 11.07 km b 31.5 km c 908 m d 2659 m e 28 721 m

6 Calculate the perimeter of each shape.

a
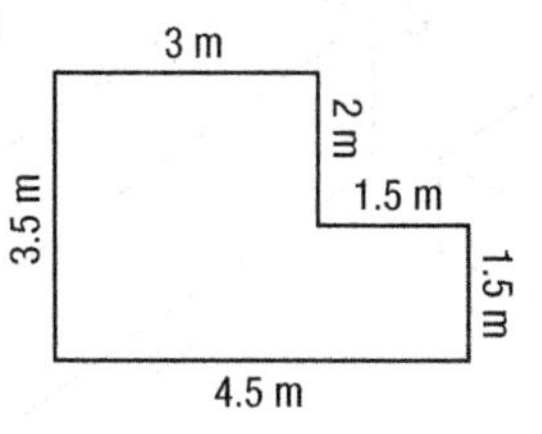

b
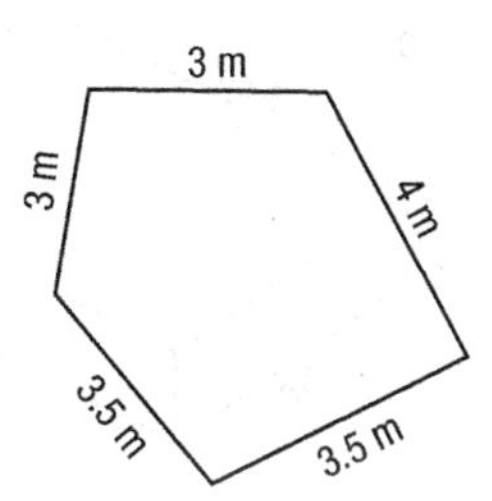

c
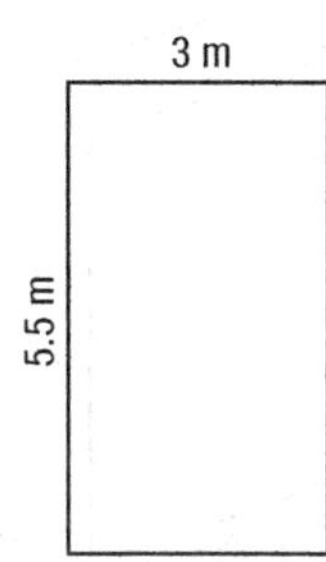

d
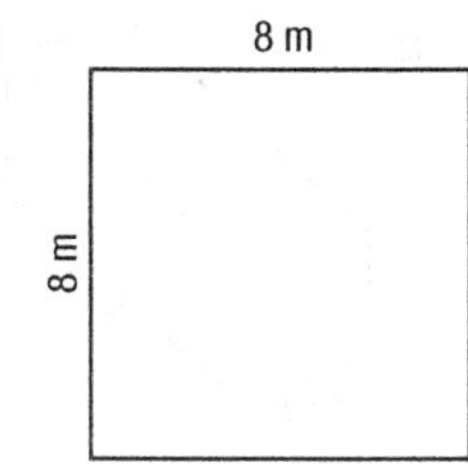

7 Calculate the area of each rectangle.

a
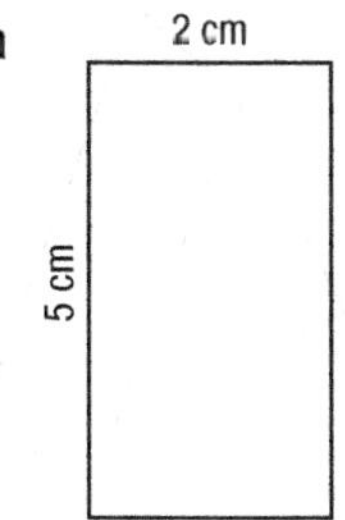

b
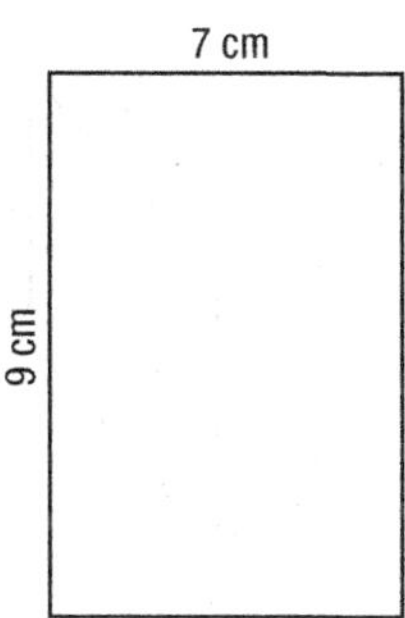

c
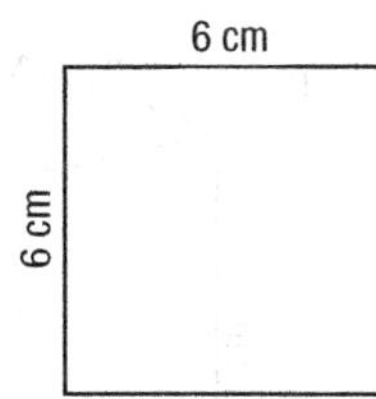

d
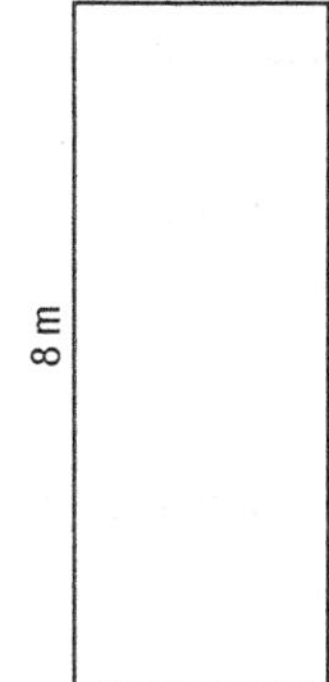

8 Calculate the area of each triangle.

a
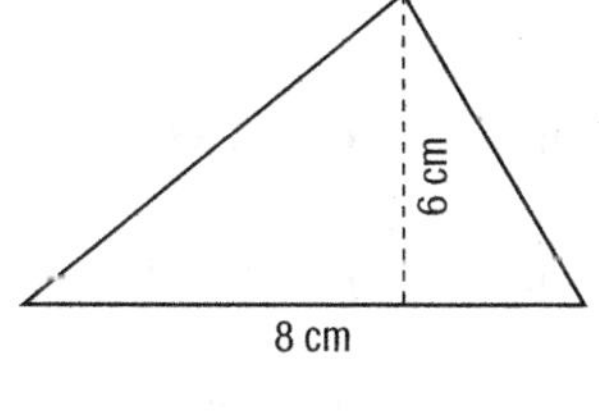

b
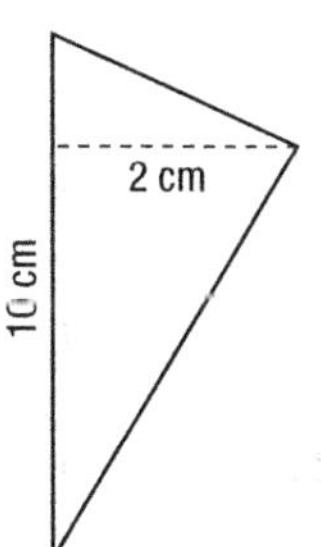

c
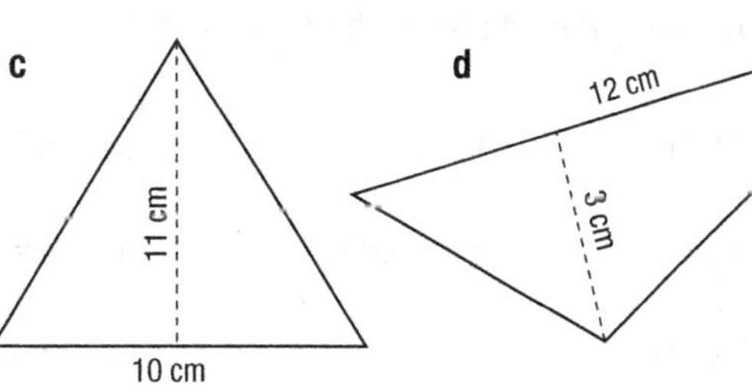

d

9 Calculate the shaded area of each shape.

a
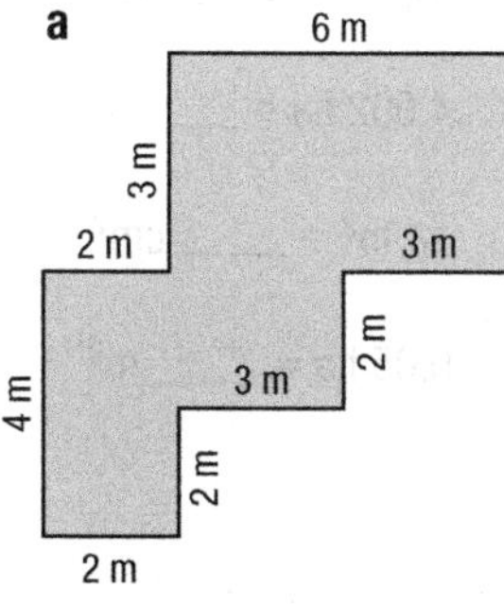

b
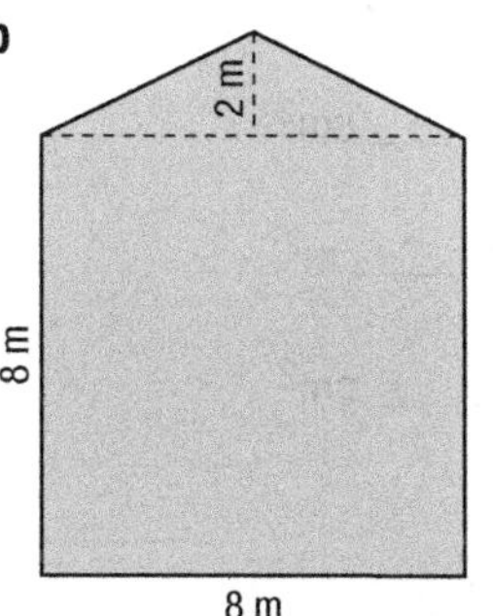

c
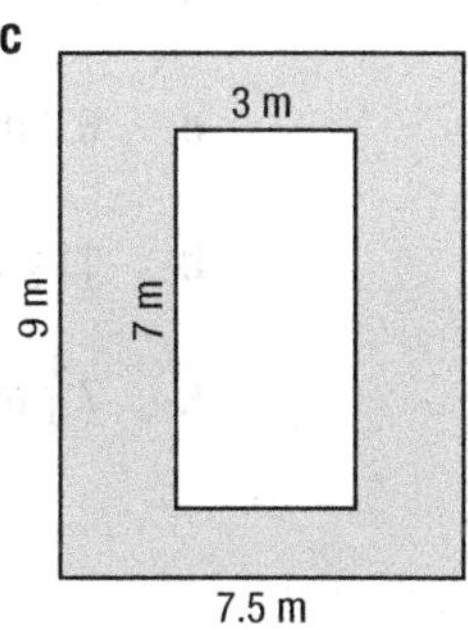

d
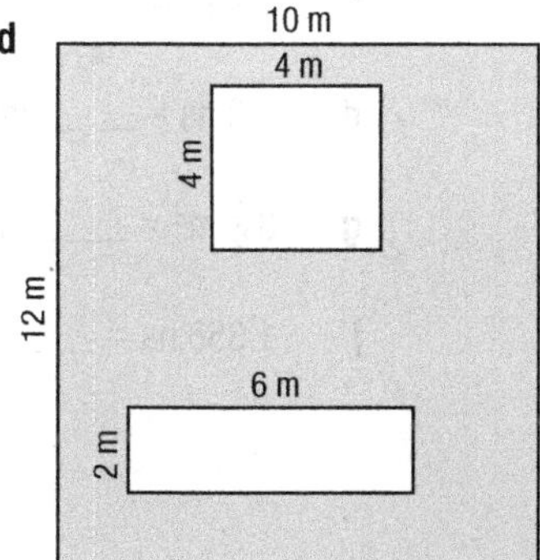

10 Calculate the volume of each rectangular prism.

a

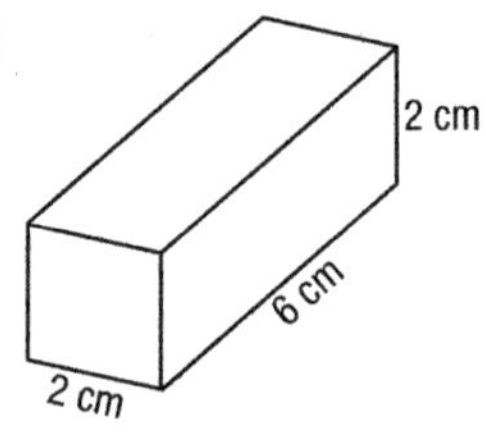

b

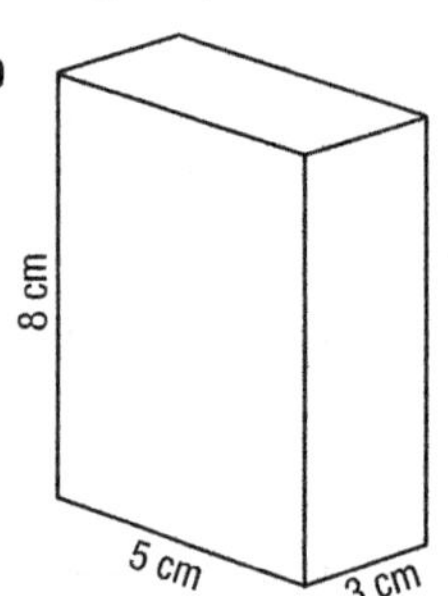

c

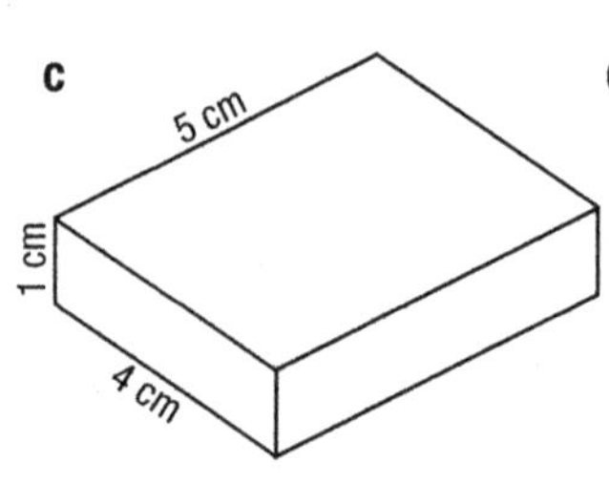

d

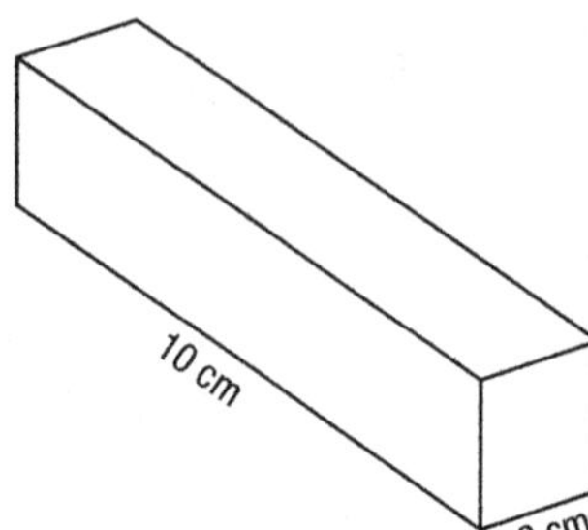

11 Calculate the volume of each triangular prism.

a

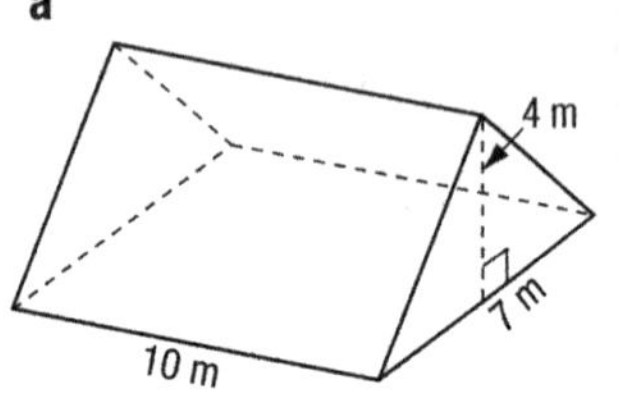

b

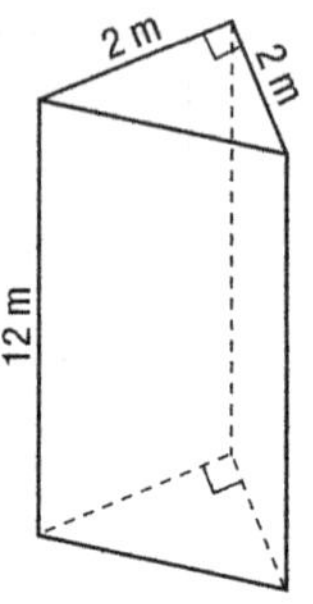

c

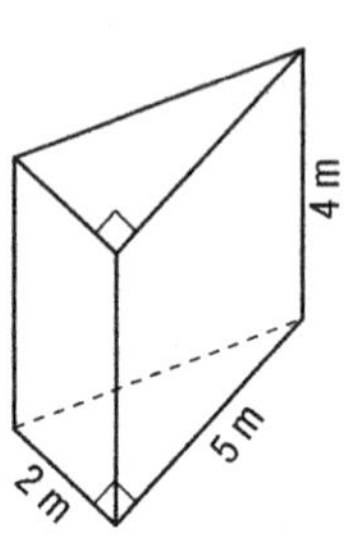

d

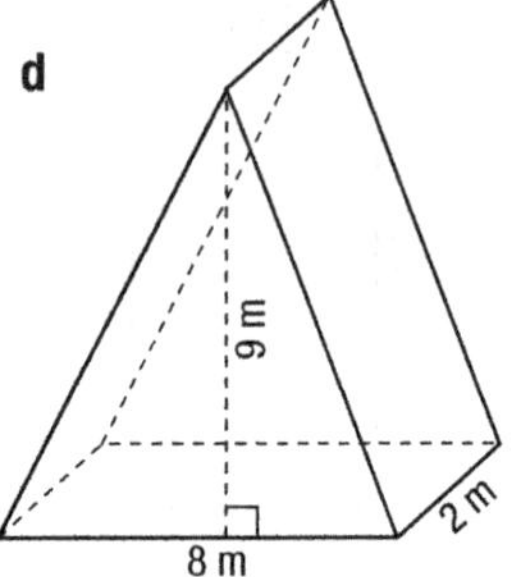

12 How many millilitres in each of these quantities?

a 8 L 265 mL **b** $7\frac{3}{4}$ L **c** 5.2 L **d** 2 L 15 mL **e** 0.86 L

f $3\frac{7}{8}$ L **g** 6.05 L **h** $4\frac{2}{5}$ L **i** 10 L 8 mL **j** 9.34 L

k 5 L 350 mL **l** $1\frac{9}{10}$ L **m** 3.422 L **n** $7\frac{1}{8}$ L **o** 18 L

13 Complete the gaps with the correct area unit.

a $10\text{ m}^2 =$ _____ cm^2 **b** 5 ha = _____ m^2 **c** $2\frac{3}{4}\text{ m}^2 =$ _____ cm^2

d 7.3 ha = _____ m^2 **e** $6.8\text{ m}^2 =$ _____ cm^2 **f** 4.002 ha = _____ m^2

g $3\frac{1}{8}\text{ m}^2 =$ _____ cm^2 **h** $11\frac{7}{8}$ ha = _____ m^2 **i** $\frac{3}{10}\text{ m}^2 =$ _____ cm^2

j 1.355 ha = _____ m^2 **k** $7\frac{4}{5}\text{ m}^2 =$ _____ cm^2 **l** 0.05 ha = _____ m^2

14 Write the name of each angle.

a 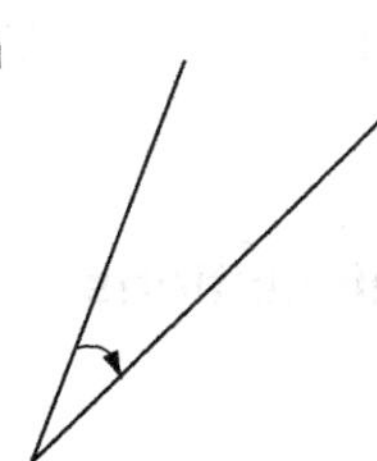

b 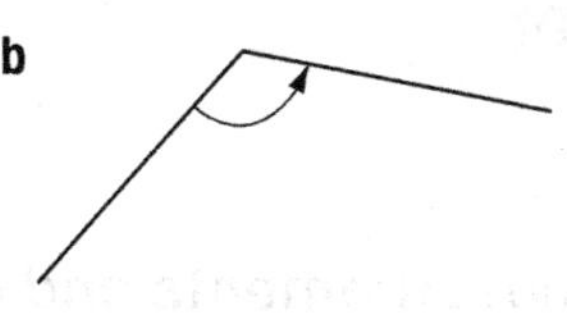

c 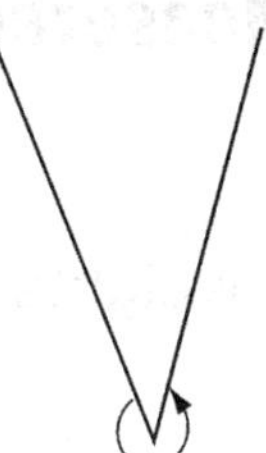

d

e 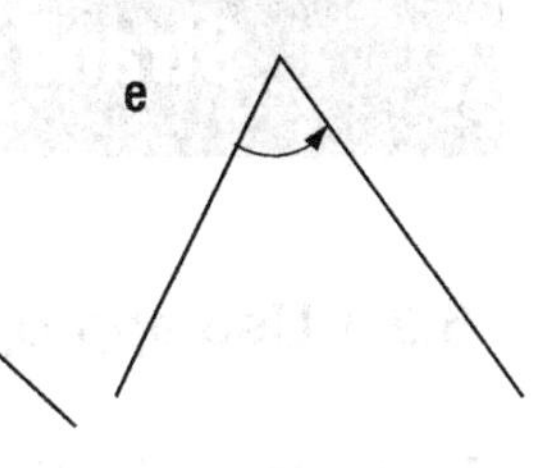

15 Use a protractor to measure each marked angle.

a 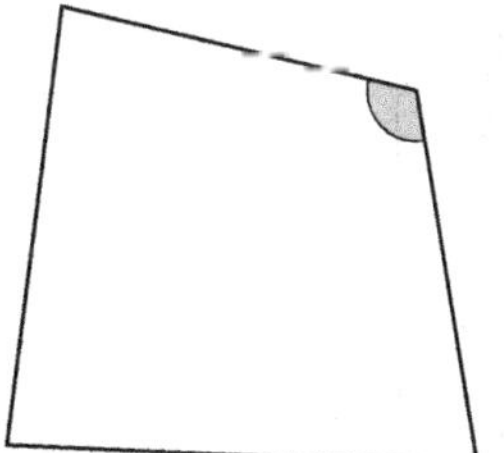

b 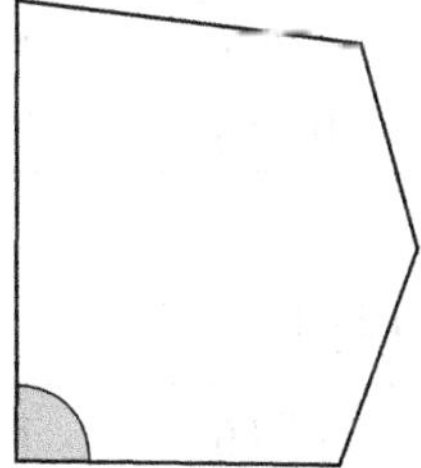

c 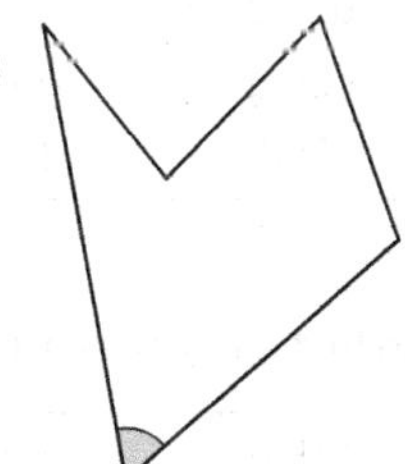

d 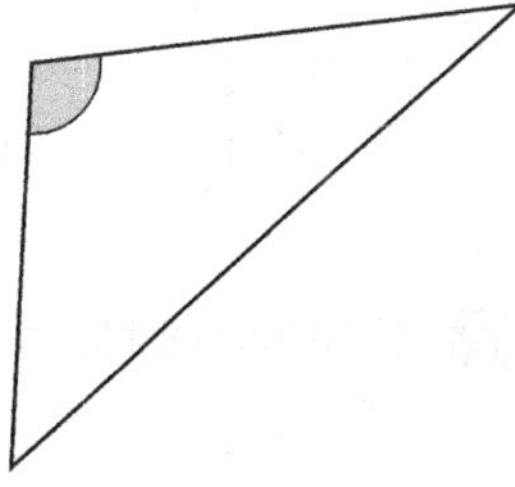

16 Calculate the missing angles.

a 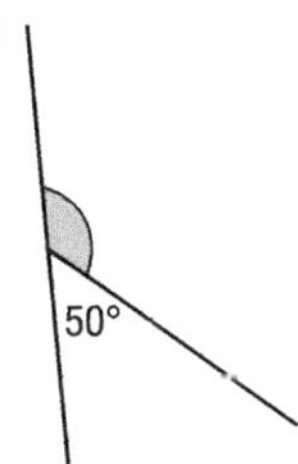

b 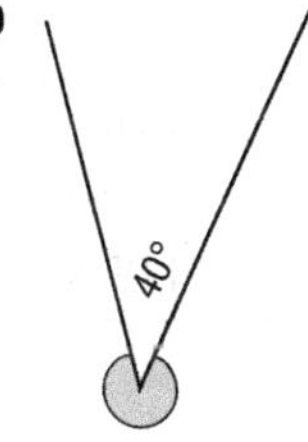

c 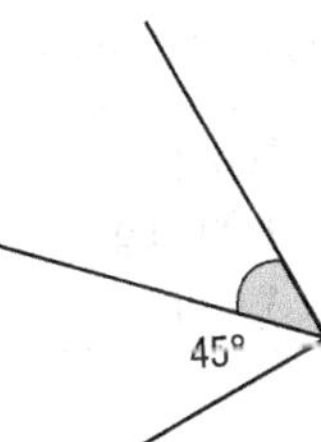

d 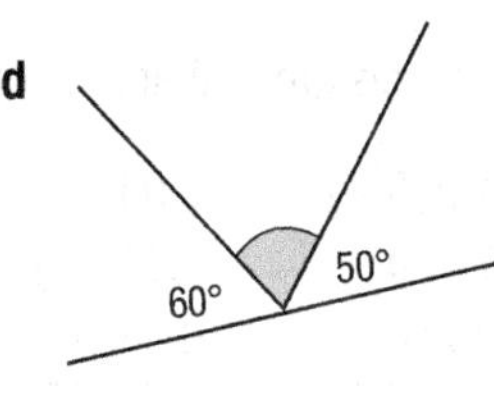

e 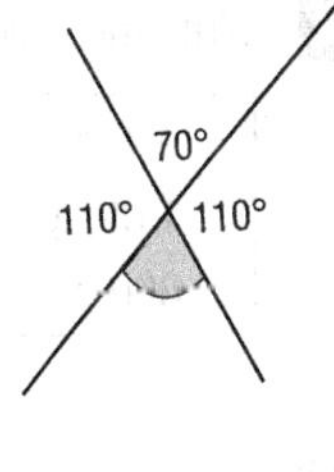

17 If the scale used on a map is 1 cm = 150 m, how many centimetres would represent these distances?

a	450 m	**b**	75 m	**c**	1200 m	**d**	1500 m	**e**	750 m	**f**	375 m
g	225 m	**h**	675 m	**i**	3000 m	**j**	1125 m	**k**	525 m	**l**	1350 m
m	300 m	**n**	1875 m	**o**	1425 m	**p**	825 m	**q**	1050 m	**r**	$2\frac{1}{4}$ km

18 If the scale used on a map is 1.5 cm = 120 km, what distances do these lengths represent?

a	3 cm	**b**	15 cm	**c**	0.75 cm	**d**	0.5 cm	**e**	3.5 cm	**f**	7.5 cm
g	1 cm	**h**	9.5 cm	**i**	12 cm	**j**	6.75 cm	**k**	5.5 cm	**l**	2 cm
m	6 cm	**n**	13 cm	**o**	30 cm	**p**	10 cm	**q**	4.5 cm	**r**	16.5 cm

Strand Measurement

6.3.1 Use appropriate units for weights in measurements and calculations

Compare and order weights

1 Write the weights from this group that are between $\frac{1}{5}$ kg and $\frac{7}{10}$ kg.

0.15 kg	0.8 kg	0.089 kg	315 g	0.25 kg	115 g
3.5 kg	710 g	600 g	0.75 kg	570 g	1.5 kg
275 g	0.68 kg	1200 g	375 g	0.5 kg	1.04 kg

2 Write the weights from this group that are between $1\frac{3}{4}$ kg and $2\frac{2}{5}$ kg.

1.65 kg	7000 g	115 g	2.06 kg	1.8 kg	1.09 kg
1250 g	0.89 kg	2.2 kg	2375 g	1.6 kg	3.6 kg
1990 g	2.04 kg	1.95 kg	2500 g	2250 g	1.76 kg

3 Write the weights from this group that are between $\frac{4}{5}$ kg and $1\frac{1}{4}$ kg.

950 g	1.275 kg	0.9 kg	805 g	1.75 kg	1 kg
1.09 kg	795 g	1.245 kg	1.035 kg	110 g	2.5 kg
860 g	450 g	0.85 kg	1.12 kg	1.23 kg	1.29 kg

4 Write the weights from this group that are between $\frac{3}{10}$ kg and 0.35 kg.

310 g	0.275 kg	0.4 kg	315 g	0.03 kg	0.5 kg
3.1 kg	400 g	3 kg	0.316 kg	230 g	0.003 kg
360 g	35 g	1.35 kg	0.33 kg	3.5 kg	3500 g

5 Record the largest weight in each group.

a 0.07 kg 0.7 kg 7 g

b 150 g 1.5 kg 1.05 kg

c 405 g 0.4 kg 0.14 kg

d 3.1 kg 3001 g 3.01 kg

e 2.75 kg 2000 g 2.8 kg

f 0.8 kg 815 g $\frac{3}{4}$ kg

g $\frac{1}{2}$ kg 0.55 kg 150 g

h 215 g $\frac{1}{4}$ kg 0.275 kg

i 675 g 0.06 kg 0.67 kg

j 0.38 kg 308 g 3800 g

k 0.75 kg 755 g $\frac{7}{10}$ kg

l 595 g 0.6 kg 0.59 kg

m 1.1 kg 1.01 kg 1050 g

n $\frac{1}{4}$ kg 0.205 kg 230 g

o 4150 g 4.05 kg 4.105 kg

p $\frac{7}{10}$ kg 710 g 0.07 kg

q 80 g 0.085 kg 0.8 kg

r 920 g 0.9 kg 0.95 kg

6 Write weights that fit these criteria:

- **a** Three weights that are greater than 0.75 kg and less than 1 kg.
- **b** Three weights that are between $\frac{3}{5}$ kg and $\frac{4}{5}$ kg.
- **c** Four weights that are greater than $\frac{1}{10}$ kg and less than $\frac{1}{5}$ kg.
- **d** Four weights that are less than 0.1 kg.
- **e** Three weights that are between 0.8 kg and 0.85 kg.
- **f** Four weights that are greater than $1\frac{1}{2}$ kg and less than 1.55 kg.

Convert between metric units of weight

1 How many grams (g) in these kilograms (kg)?

a	1.5 kg	**b**	2.75 kg	**c**	0.695 kg	**d**	1.2 kg	**e**	4.35 kg	**f**	0.636 kg
g	0.178 kg	**h**	3.08 kg	**i**	2.195 kg	**j**	1.056 kg	**k**	5.95 kg	**l**	0.372 kg
m	5.07 kg	**n**	0.005 kg	**o**	2.106 kg	**p**	1.09 kg	**q**	9.12 kg	**r**	5.97 kg
s	3.1 kg	**t**	0.087 kg	**u**	2.05 kg	**v**	0.441 kg	**w**	1.703 kg	**x**	3.79 kg

2 Use decimal notation where necessary to write these as kilograms (kg).

a	2000 g	**b**	1500 g	**c**	7000 g	**d**	5500 g	**e**	3250 g	**f**	1750 g
g	4250 g	**h**	8750 g	**i**	6700 g	**j**	1850 g	**k**	1520 g	**l**	2780 g
m	850 g	**n**	590 g	**o**	3054 g	**p**	7258 g	**q**	1067 g	**r**	609 g
s	54 g	**t**	5005 g	**u**	2320 g	**v**	13 081 g	**w**	15 690 g	**x**	10 700 g

3 How many kilograms (kg) in these tonnes (t)?

a	7 t	**b**	11 t	**c**	4.5 t	**d**	6.25 t	**e**	3.25 t	**f**	7.05 t
g	15.1 t	**h**	3.067 t	**i**	2.4 t	**j**	0.134 t	**k**	5.75 t	**l**	8.7 t
m	0.12 t	**n**	1.088 t	**o**	4.75 t	**p**	3.875 t	**q**	6.97 t	**r**	0.004 t
s	2.096 t	**t**	12.4 t	**u**	10.07 t	**v**	7.058 t	**w**	0.099 t	**x**	1.53 t

4 Use decimal notation where necessary to write these as tonnes (t).

a	9000 kg	**b**	2000 kg	**c**	13 000 kg	**d**	21 000 kg	**e**	6500 kg	**f**	4500 kg
g	9250 kg	**h**	3750 kg	**i**	5268 kg	**j**	1752 kg	**k**	2085 kg	**l**	674 kg
m	8600 kg	**n**	9107 kg	**o**	510 kg	**p**	3583 kg	**q**	4590 kg	**r**	10 070 kg
s	1316 kg	**t**	830 kg	**u**	12 158 kg	**v**	1316 kg	**w**	380 kg	**x**	75 kg

5 Complete the gaps.

a	$2\frac{1}{4}$ kg = ____ g	b	$5\frac{3}{4}$ t = ____ kg	c	$3\frac{1}{2}$ kg = ____ g	d	$7\frac{1}{4}$ t = ____ kg
e	$\frac{1}{2}$ kg = ____ g	f	$1\frac{1}{2}$ t = ____ kg	g	$\frac{1}{5}$ kg = ____ g	h	$3\frac{2}{5}$ t = ____ kg
i	$8\frac{3}{10}$ kg = ____ g	j	$\frac{3}{5}$ t = ____ kg	k	$5\frac{7}{10}$ kg = ____ g	l	$4\frac{1}{8}$ t = ____ kg
m	$2\frac{5}{8}$ kg = ____ g	n	$6\frac{3}{8}$ t = ____ kg	o	$4\frac{7}{8}$ kg = ____ g	p	$\frac{7}{8}$ t = ____ kg
q	$7\frac{1}{8}$ kg = ____ g	r	$1\frac{1}{20}$ t = ____ kg	s	$9\frac{11}{20}$ kg = ____ g	t	$4\frac{9}{25}$ t = ____ kg

Solve weight calculations

1 Write answers to the following in kilograms (kg).

a	$\frac{1}{4}$ of 1 t	b	$\frac{3}{8}$ of 2 t	c	$\frac{3}{4}$ of 2 t	d	$\frac{1}{8}$ of 4 t	e	$\frac{2}{5}$ of 2 t	f	$\frac{4}{5}$ of 3 t
g	$\frac{7}{10}$ of 5 t	h	$\frac{1}{5}$ of 1 t	i	$\frac{5}{8}$ of 3 t	j	$\frac{11}{20}$ of 2 t	k	$\frac{4}{25}$ of 1 t	l	$\frac{17}{25}$ of 3 t

2 Write answers to the following in grams (g).

a	10% of 2 kg	b	25% of 3 kg	c	50% of 1 kg	d	20% of 2 kg
e	5% of 4 kg	f	75% of 5 kg	g	10% of 3 kg	h	25% of 5 kg
i	20% of 7 kg	j	40% of 3 kg	k	80% of 2 kg	l	75% of 3 kg

3 Write answers to the following in tonnes (t), using decimal notation where necessary.

a	300 kg × 8	b	250 kg × 12	c	420 kg × 7	d	180 kg × 15
e	500 kg × 10	f	740 kg × 5	g	650 kg × 20	h	810 kg × 6
i	490 kg × 11	j	370 kg × 25	k	290 kg × 12	l	705 kg × 9

4 Write answers to the following in kilograms (kg), using decimal notation where necessary.

a	400 g × 8	b	550 g × 6	c	375 g × 5	d	725 g × 9
e	235 g × 7	f	895 g × 4	g	645 g × 11	h	575 g × 8
i	165 g × 30	j	915 g × 20	k	285 g × 15	l	455 g × 12

5 Convert the different weights to the same unit of measurement and calculate.

a 6.5 kg + 1300 g + 1.05 kg

b $2\frac{3}{4}$ kg + 0.8 kg + 480 g

c 3.5 t – 850 kg

d 7.25 kg – 845 g

e $4\frac{3}{4}$ t + 0.65 t + 917 kg

f 475 kg + 1.04 t + 790 kg

g 11.6 kg – 1275 g

h 2.15 t – 865 kg

i 780 g + 1525 g + 2.4 kg

j 1.625 kg + 495 g + 0.03 kg

k 0.75 t – 287 kg

l 9.2 kg – 2567 g

6 Copy and complete the following chart.

	Total weight	Number of items of the same weight	Weight of one item
a	1.775 kg	5	
b	6.3 t	6	
c	54.6 kg	7	
d		9	725 g
e		4	2.85 kg
f	0.95 t	8	
g		7	5.27 kg
h	3.892 kg	4	

6.3.3 Investigate temperature and the thermometer as a measuring instrument

Read thermometer scales

What temperature is shown on each thermometer?

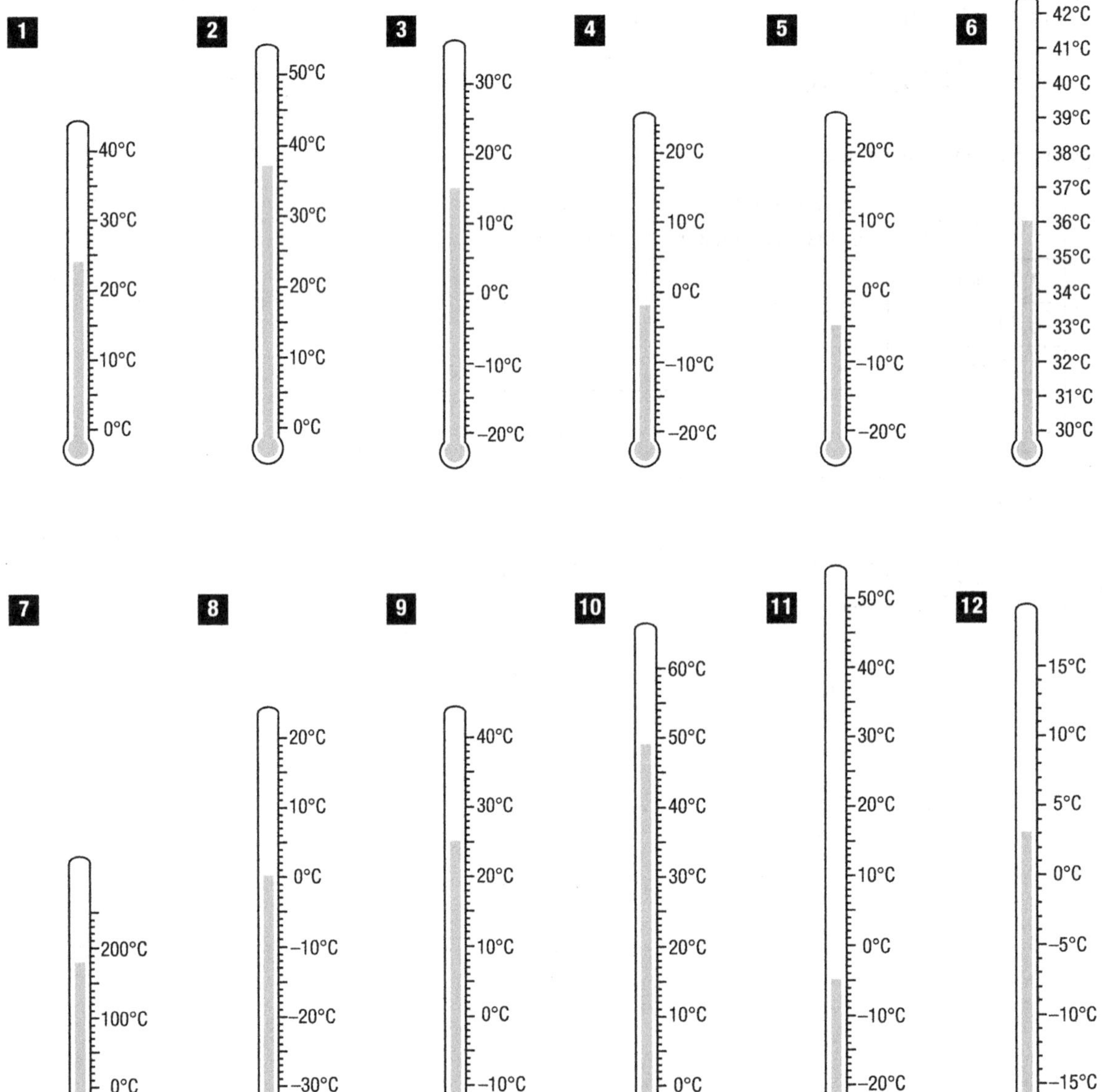

6.3.4 Use a variety of time records

Read and record time to the nearest minute

Write the time from each clock face in analogue and digital form; for example, the first clock shows $\frac{1}{2}$ past 6 or 6:30.

1

2

3

4

5

6

7

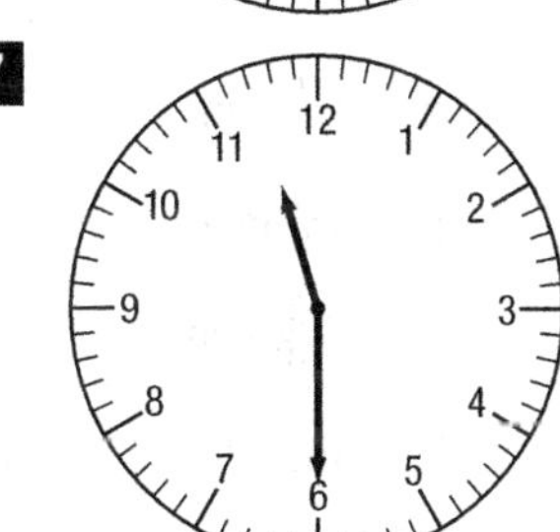

8

9

10

11

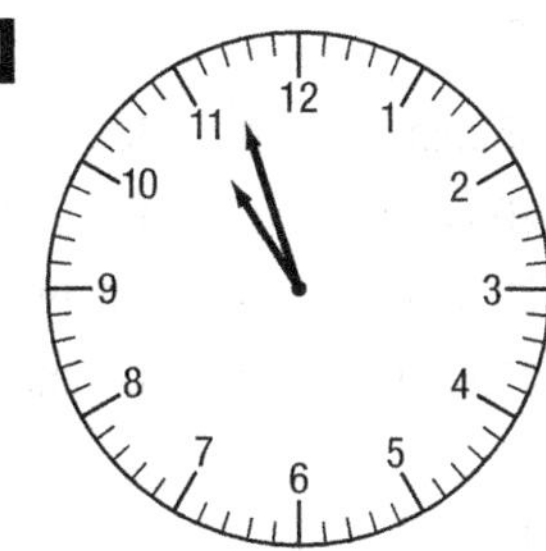

12

13

14

15

Convert between time units

1 How many minutes in these hours?

a	2 hours	**b**	5 hours	**c**	3 hours	**d**	10 hours	**e**	$1\frac{1}{2}$ hours
f	$6\frac{1}{2}$ hours	**g**	$2\frac{1}{4}$ hours	**h**	$8\frac{1}{4}$ hours	**i**	$3\frac{3}{4}$ hours	**j**	$1\frac{3}{4}$ hours
k	$9\frac{3}{4}$ hours	**l**	$5\frac{3}{4}$ hours	**m**	$2\frac{1}{3}$ hours	**n**	$6\frac{1}{3}$ hours	**o**	$\frac{2}{3}$ hour
p	$4\frac{2}{3}$ hours	**q**	$\frac{1}{5}$ hour	**r**	$7\frac{1}{5}$ hours	**s**	$5\frac{3}{5}$ hours	**t**	$1\frac{4}{5}$ hours
u	$8\frac{1}{10}$ hours	**v**	$\frac{9}{10}$ hour	**w**	$2\frac{7}{10}$ hours	**x**	$4\frac{3}{10}$ hours		

2 Write these hours (h) and minutes (min) as minutes.

a	5 h 25 min	**b**	3 h 10 min	**c**	8 h 45 min	**d**	2 h 50 min	**e**	1 h 35 min
f	4 h 22 min	**g**	9 h 18 min	**h**	3 h 46 min	**i**	7 h 19 min	**j**	1 h 57 min
k	2 h 7 min	**l**	5 h 32 min	**m**	3 h 24 min	**n**	10 h 41 min	**o**	6 h 53 min
p	4 h 49 min	**q**	8 h 11 min	**r**	5 h 27 min	**s**	7 h 14 min	**t**	1 h 26 min
u	2 h 39 min	**v**	4 h 3 min	**w**	11 h 43 min	**x**	6 h 21 min		

3 How many seconds in these minutes?

a	12 min	**b**	4 min	**c**	$3\frac{1}{2}$ min	**d**	$7\frac{1}{2}$ min	**e**	$2\frac{1}{4}$ min
f	$5\frac{3}{4}$ min	**g**	$10\frac{1}{3}$ min	**h**	$8\frac{2}{3}$ min	**i**	$4\frac{2}{5}$ min	**j**	$7\frac{4}{5}$ min
k	$1\frac{3}{5}$ min	**l**	$6\frac{1}{5}$ min	**m**	$3\frac{1}{10}$ min	**n**	$8\frac{9}{10}$ min	**o**	$5\frac{3}{10}$ min
p	$2\frac{7}{10}$ min	**q**	$12\frac{1}{4}$ min	**r**	$11\frac{3}{4}$ min	**s**	$4\frac{1}{6}$ min	**t**	$6\frac{5}{6}$ min

4 Write these minutes (min) as hours and minutes.

a	245 min	**b**	130 min	**c**	85 min	**d**	320 min	**e**	500 min	**f**	108 min
g	146 min	**h**	197 min	**i**	205 min	**j**	413 min	**k**	311 min	**l**	269 min
m	624 min	**n**	517 min	**o**	73 min	**p**	398 min	**q**	456 min	**r**	552 min
s	174 min	**t**	226 min	**u**	339 min	**v**	431 min	**w**	508 min	**x**	587 min

5 Write these seconds (s) as minutes and seconds.

a	225 s	**b**	94 s	**c**	312 s	**d**	566 s	**e**	113 s	**f**	286 s
g	461 s	**h**	389 s	**i**	76 s	**j**	158 s	**k**	251 s	**l**	195 s
m	343 s	**n**	502 s	**o**	410 s	**p**	128 s	**q**	219 s	**r**	396 s
s	104 s	**t**	627 s	**u**	826 s	**v**	81 s	**w**	263 s	**x**	482 s

6 How many hours (h) in these days (d)?

a 5 d b 2 d c 7 d d 4 d e 3 d and 2 h
f 2 d and 11 h g 5 d and 17 h h 6 d and 5 h i 10 d and 4 h j 8 d and 13 h
k 3 d and 9 h l 7 d and 8 h m $2\frac{1}{4}$ d n $6\frac{1}{2}$ d o $5\frac{1}{3}$ d
p $3\frac{3}{4}$ d q $4\frac{2}{3}$ d r $7\frac{1}{8}$ d s $3\frac{5}{8}$ d t $11\frac{1}{2}$ d

7 How many days in these weeks?

a 6 weeks b 4 weeks and 2 days c 3 weeks and 4 days
d 8 weeks and 5 days e 6 weeks and 1 day f 5 weeks and 3 days
g 10 weeks and 2 days h 2 weeks and 6 days i 7 weeks and 5 days
j 4 weeks and 6 days k 3 weeks and 2 days l 12 weeks
m 15 weeks n 22 weeks o 19 weeks
p 11 weeks and 3 days q 12 weeks and 6 days r 17 weeks and 3 days

8 Complete the gaps.

a 2 years = _____ months b 3 decades = _____ years c 5 years = _____ months
d 3 centuries = _____ years e 8 years = _____ months f 6 decades = _____ years
g $4\frac{1}{2}$ years = _____ months h 7 centuries = _____ years i $2\frac{1}{4}$ years = _____ months
j $2\frac{1}{2}$ centuries = _____ years k $5\frac{1}{2}$ decades = _____ years l $3\frac{3}{4}$ years = _____ months
m $1\frac{1}{2}$ decades = _____ years n $4\frac{1}{4}$ centuries = _____ years o $2\frac{1}{5}$ decades = _____ years
p $6\frac{2}{3}$ years = _____ months q $9\frac{3}{5}$ decades = _____ years r $4\frac{1}{3}$ years = _____ months
s 2 centuries = _____ decades t $3\frac{1}{2}$ centuries = _____ decades u $1\frac{1}{2}$ centuries = _____ decades

9 Write these hours as days and hours, using fractions where necessary.

a 48 h b 12 h c 96 h d 240 h e 6 h f 36 h
g 72 h h 84 h i 30 h j 120 h k 18 h l 60 h
m 27 h n 54 h o 16 h p 56 h q 104 h r 160 h

10 Write these months as years and months, using fractions where necessary.

a 24 months b 60 months c 6 months d 72 months e 18 months
f 36 months g 78 months h 16 months i 32 months j 51 months
k 105 months l 63 months m 56 months n 112 months o 144 months
p 90 months q 40 months r 33 months s 123 months t 15 months

Convert between 12-hour and 24-hour clock time

1 Write these 24-hour clock times as 12-hour clock times, for example: 1520 = 3:20 p.m.

a	1320	**b**	2043	**c**	1619	**d**	0827	**e**	1954	**f**	1411
g	0552	**h**	1738	**i**	2205	**j**	0942	**k**	1056	**l**	2316
m	1432	**n**	0734	**o**	1510	**p**	0900	**q**	1706	**r**	1345
s	0241	**t**	1859	**u**	2027	**v**	0312	**w**	2139	**x**	1917

2 Write these 12-hour clock times as 24-hour clock times.

a	2:40 p.m.	**b**	5:24 p.m.	**c**	11:47 a.m.	**d**	10:06 p.m.	**e**	7 p.m.
f	6:39 a.m.	**g**	8:58 p.m.	**h**	3:16 p.m.	**i**	4:42 a.m.	**j**	6:21 p.m.
k	2:13 p.m.	**l**	12:08 p.m.	**m**	10 a.m.	**n**	12:15 p.m.	**o**	7:45 a.m.
p	1:55 p.m.	**q**	9:36 p.m.	**r**	5:50 a.m.	**s**	10:52 p.m.	**t**	6:05 a.m.
u	11:59 p.m.	**v**	12:45 a.m.	**w**	4 p.m.	**x**	3:30 p.m.		

3

1550	0859	2205	1735	2023	1412	1148	1815

a Which times are after 6 p.m.?

b Which times are before 5 p.m.?

c Which time is nearest to 9 a.m.?

d Which time is closest to midnight?

e Which times are between 3 p.m. and 6 p.m.?

4 Copy this table and complete the missing parts.

	Morning/afternoon	Clock	12-hour time	24-hour time
a	Morning			
b			10:40 p.m.	
c				1722
d	Afternoon			
e			1.34 p.m.	

Calculate with 12-hour and 24-hour time

1 How many minutes does each PMV take to reach its destination?

	Depart	Arrive		Depart	Arrive
a	1035	1054	**b**	0957	1021
c	0846	0938	**d**	1103	1145
e	1026	1104	**f**	0946	1012
g	1117	1151	**h**	1149	1213
i	1335	1356	**j**	1516	1544
k	1053	1124	**l**	1358	1423
m	0715	0806	**n**	0648	0723
o	1637	1712	**p**	1446	1508
q	1819	1855	**r**	1254	1336
s	1552	1641	**t**	1705	1732

2 If each journey is 47 minutes, calculate the arrival time for each PMV if they depart at the following times.

a	6:30 p.m.	**b**	4:12 p.m.	**c**	9:52 a.m.	**d**	11:27 a.m.	**e**	2:45 p.m.
f	1:16 p.m.	**g**	10:05 a.m.	**h**	3:38 p.m.	**i**	2:19 p.m.	**j**	11:42 a.m.
k	9:23 a.m.	**l**	7:48 p.m.	**m**	5:11 p.m.	**n**	8:39 a.m.	**o**	6:53 a.m.
p	5:17 a.m.	**q**	10:41 a.m.	**r**	4:59 p.m.	**s**	8:07 p.m.	**t**	9:13 p.m.
u	11:06 a.m.	**v**	1:45 p.m.	**w**	12:33 p.m.	**x**	6:24 a.m.		

3 If each game took 38 minutes, use the following finish times to calculate the starting time for each game.

a	2:30 p.m.	**b**	10:10 a.m.	**c**	11:40 a.m.	**d**	3:10 p.m.	**e**	1:15 p.m.
f	2:06 p.m.	**g**	10:32 a.m.	**h**	12:26 p.m.	**i**	4:08 p.m.	**j**	1:24 p.m.
k	5:12 p.m.	**l**	3:18 p.m.	**m**	9:45 a.m.	**n**	10:55 a.m.	**o**	12:38 p.m.
p	4:50 p.m.	**q**	5:05 p.m.	**r**	2:47 p.m.	**s**	11:12 a.m.	**t**	1:29 p.m.
u	8:50 a.m.	**v**	12:51 p.m.	**w**	4:22 p.m.	**x**	9:07 a.m.		

4 Use these starting and finishing times to calculate the length of time each person worked.

	Start	Finish		Start	Finish
a	11:10 a.m.	4:45 p.m.	**b**	8:50 a.m.	2:25 p.m.
c	9:30 a.m.	3:50 p.m.	**d**	8:25 a.m.	2:55 p.m.
e	7:15 a.m.	12:40 p.m.	**f**	10:40 a.m.	2:30 p.m.
g	6:20 a.m.	11:50 a.m.	**h**	1:30 p.m.	6:05 p.m.
i	12:45 p.m.	6:50 p.m.	**j**	1:35 p.m.	5:10 p.m.
k	2 p.m.	8:55 p.m.	**l**	4:15 p.m.	10:40 p.m.
m	7:45 a.m.	1:30 p.m.	**n**	8:10 a.m.	3:30 p.m.
o	5:40 p.m.	11:50 p.m.	**p**	3:30 p.m.	9:55 p.m.
q	5:05 a.m.	10:45 a.m.	**r**	6:45 a.m.	1:50 p.m.

Calculate with calendar time

1 Which of these children will be under 14 years of age on 1st November 2018?

Name	*Date of birth*	*Name*	*Date of birth*
Simon	16/09/2004	Ruby	03/12/2004
Naomi	27/10/2004	Karu	05/01/2005
Tagu	06/07/2004	Zekele	31/12/2003
Rani	04/02/2005	Michael	23/08/2004
Betty	18/10/2004	Lucy	06/11/2004
John	20/01/2005	Tovorika	15/06/2004
Ruth	11/03/2003	Sam	20/09/2005

2 Which of these children will be under 12 years of age on 30th June 2020?

Name	*Date of birth*	*Name*	*Date of birth*
Peter	04/03/2008	Tau	17/09/2008
Pipini	15/05/2008	Ligo	29/06/2008
Elli	25/10/2008	Renaki	06/11/2008
Joseph	31/07/2008	David	10/04/2008
Mary	02/01/2009	Charles	13/09/2008
Julius	21/07/2008	Mirou	12/01/2009
Paul	07/02/2008	Vagi	21/03/2008

3 Which of these children will be under 10 years of age on 15th April 2017?

Name	*Date of birth*	*Name*	*Date of birth*
Steven	19/06/2007	Otto	03/04/2007
Kila	25/04/2007	Dorcas	24/03/2007
Manu	26/09/2007	Bernadette	17/08/2007
Josephine	10/01/2007	Nalda	05/01/2008
Annette	14/05/2007	Saki	18/03/2007
Fabian	11/04/2007	Ryan	30/07/2007
Stanley	09/02/2008	Edward	11/09/2007

4 Which of these children will be under 16 years of age on 10th January 2019?

Name	*Date of birth*	*Name*	*Date of birth*
Alena	09/04/2003	Phillip	24/12/2002
Pulu	13/02/2003	Aquala	05/01/2003
Kari	18/06/2003	Abai	22/11/2002
Sophie	06/07/2003	Matmilo	21/12/2002
Lemeck	01/01/2003	Tapo	15/01/2003
Freda	14/10/2002	Amos	27/03/2003
Nora	21/04/2003	Jerry	23/05/2003

Assessment Measurement

1 Put each set of weights in order from heaviest to lightest.

a 0.8 kg 0.08 kg 8 kg **b** 280 g 0.2 kg 0.25 kg **c** 5.6 t $5\frac{3}{4}$ t 5500 kg

d 4.75 t 4780 kg 4.075 t **e** $2\frac{3}{4}$ kg 2.7 kg 2900 g **f** 3125 kg 3.01 t $3\frac{3}{8}$ t

g 0.04 t 50 kg 0.4 t **h** $2\frac{3}{5}$ t 2.06 t 2660 kg **i** 7.1 kg 710 g $7\frac{3}{10}$ kg

j 1.5 kg 1050 g $1\frac{2}{5}$ kg

2 How many grams in these kilograms?

a $\frac{4}{5}$ kg **b** 7.5 kg **c** 3.95 kg **d** $2\frac{3}{8}$ kg **e** 9.03 kg **f** $5\frac{2}{5}$ kg

g $10\frac{3}{4}$ kg **h** 0.006 kg **i** $2\frac{7}{10}$ kg **j** 11.65 kg **k** $6\frac{1}{8}$ kg **l** 1.17 kg

3 How many kilograms in these tonnes?

a $1\frac{1}{4}$ t **b** 8.2 t **c** $\frac{7}{8}$ t **d** 0.105 t **e** 4.17 t **f** $3\frac{17}{25}$ t

g 2.92 t **h** $5\frac{9}{10}$ t **i** 6.03 t **j** $7\frac{5}{8}$ t **k** 3.9 t **l** $1\frac{3}{5}$ t

4 Use decimal notation to write these weights as kilograms or tonnes.

a 5600 kg **b** 8150 g **c** 489 kg **d** 9002 g **e** 11 025 kg **f** 76 g
g 3208 kg **h** 6430 g **i** 1213 kg **j** 5810 g **k** 92 kg **l** 1050 g

5 Write answers to these in kilograms or grams.

a 50% of 1 t **b** $\frac{5}{8}$ of 2 t **c** 10% of 4 kg **d** $\frac{9}{10}$ of 3 t **e** $\frac{3}{4}$ of 5 kg

f 75% of 2 t **g** $\frac{3}{5}$ of 6 kg **h** 20% of 8 kg **i** $\frac{4}{5}$ of 1 t **j** 25% of 2 kg

k $\frac{1}{4}$ of 1 kg **l** 5% of 3 t **m** 10% of 1 kg **n** $\frac{3}{8}$ of 3 t **o** 40% of 4 t

p $\frac{7}{10}$ of 4 kg **q** $\frac{7}{8}$ of 5 t **r** 60% of 2 kg **s** $\frac{2}{5}$ of 7 t **t** 75% of 3 kg

6 Use decimal notation to write answers in tonnes or kilograms.

a	360 kg × 6	**b**	215 g × 8	**c**	540 kg × 5	**d**	435 g × 7	**e**	194 kg × 10
f	688 g × 3	**g**	273 kg × 4	**h**	152 g × 6	**i**	5.6 kg ÷ 8	**j**	3.675 t ÷ 7
k	2.448 kg ÷ 4	**l**	1.91 t ÷ 5	**m**	2.838 kg ÷ 6	**n**	7.02 t ÷ 6	**o**	8.58 kg ÷ 3
p	0.91 t ÷ 7								

7 Calculate the answers to these additions and subtractions.

a	2.25 t – 910 kg	**b**	0.6 kg + 1.45 kg + 275 g	**c**	3.16 t + 1600 kg + 1.2 t
d	5.08 kg – 634 g	**e**	1.54 kg – 789 g	**f**	425 g + 948 g + 0.77 kg
g	2.06 t + 4 t + 1060 kg	**h**	12.8 t – $5\frac{5}{8}$ t	**i**	0.85 kg – $\frac{1}{8}$ kg
j	1.76 kg + 775 g + 0.09 kg				

8 Write the time for each clock face in analogue and digital time.

a

b

c

9 How many minutes?

a	$3\frac{1}{4}$ hours	**b**	5.5 hours	**c**	2 h 20 min	**d**	$1\frac{1}{3}$ hours	**e**	0.75 hours	**f**	5 h 10 min
g	$4\frac{1}{2}$ hours	**h**	3.7 hours	**i**	1 h 55 min	**j**	$2\frac{2}{5}$ hours	**k**	1.2 hours	**l**	4 h 35 min

10 How many seconds?

a	$9\frac{1}{2}$ min	**b**	2.5 min	**c**	5 min 20 s	**d**	$4\frac{4}{5}$ min	**e**	6.75 min	**f**	3 min 50 s
g	$1\frac{1}{3}$ min	**h**	3.8 min	**i**	2 min 35 s	**j**	$7\frac{2}{3}$ min	**k**	4.1 min	**l**	6 min 5 s

11 Write these minutes as hours and minutes.

a	86 min	**b**	195 min	**c**	110 min	**d**	216 min	**e**	71 min
f	338 min	**g**	521 min	**h**	292 min	**i**	407 min	**j**	384 min

12 Complete the gaps.

- **a** 3 days = _____ hours
- **b** 9 weeks = _____ days
- **c** 4 years = _____ months
- **d** $2\frac{1}{2}$ decades = _____ years
- **e** $2\frac{7}{8}$ days = _____ hours
- **f** 6 centuries = _____ years
- **g** 8 weeks and 3 days = _____ days
- **h** $3\frac{3}{4}$ years = _____ months
- **i** $3\frac{1}{5}$ centuries = _____ decades
- **j** $7\frac{4}{5}$ decades = _____ years
- **k** $7\frac{2}{3}$ years = _____ months
- **l** 96 hours = _____ days
- **m** 84 months = _____ years
- **n** 300 years = _____ decades
- **o** 720 seconds = _____ minutes
- **p** $12\frac{1}{3}$ days = _____ hours
- **q** $5\frac{7}{10}$ centuries = _____ years
- **r** 180 months = _____ years

13 Write these 24-hour clock times as 12-hour clock times.

a	1510	**b**	0830	**c**	1828	**d**	2145	**e**	2350	**f**	0340
g	1017	**h**	1352	**i**	1705	**j**	0938	**k**	2019	**l**	1936

14 Write these 12-hour clock times as 24-hour clock times.

a	3:23 a.m.	**b**	2:41 p.m.	**c**	11:20 a.m.	**d**	7:56 a.m.	**e**	6:37 p.m.	**f**	8:09 p.m.
g	5:44 a.m.	**h**	4:18 p.m.	**i**	1:37 p.m.	**j**	9:48 p.m.	**k**	10:55 a.m.	**l**	12:04 p.m.
m	6:32 a.m.	**n**	2:53 p.m.	**o**	5:46 a.m.						

15 The clocks below show the start time of some rugby games. If a game takes 90 minutes, what time will each of these games finish?

a **b** **c** **d**

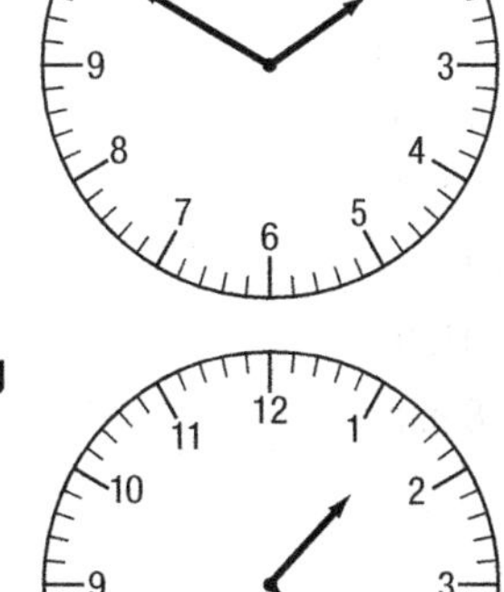

e **f** **g** **h**

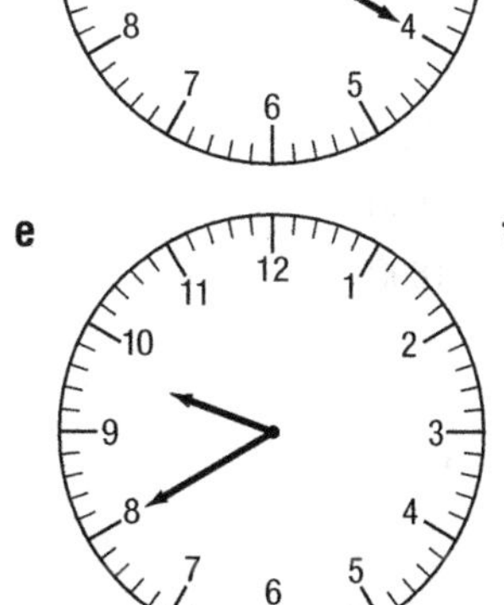

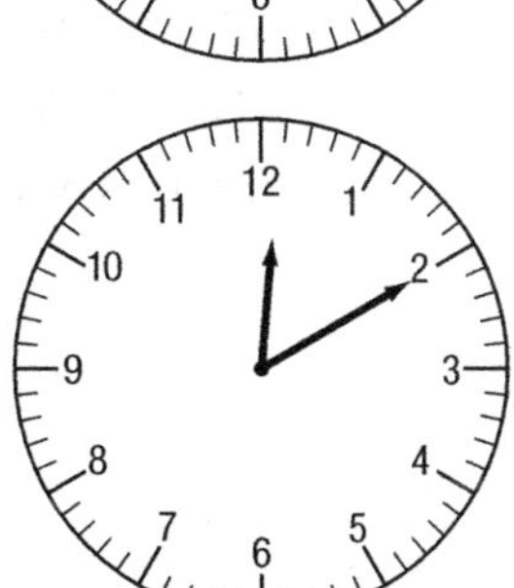

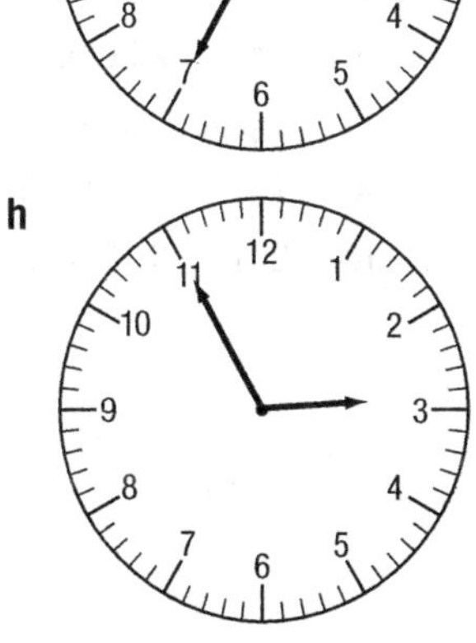

16 The clocks below show the finish times of some basketball matches. If a match takes 40 minutes, what time did each of these matches start?

a

b

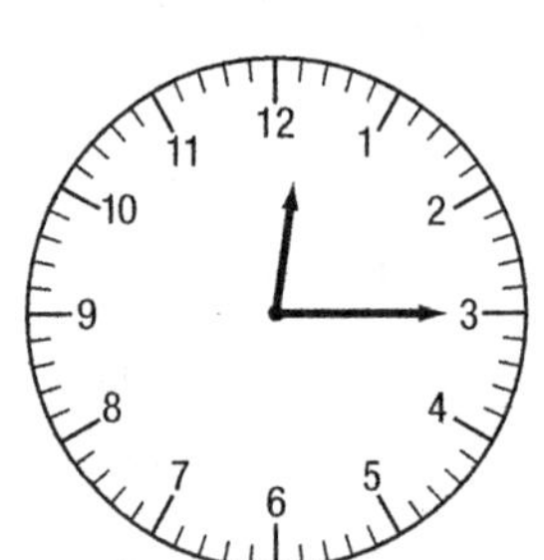

c

d

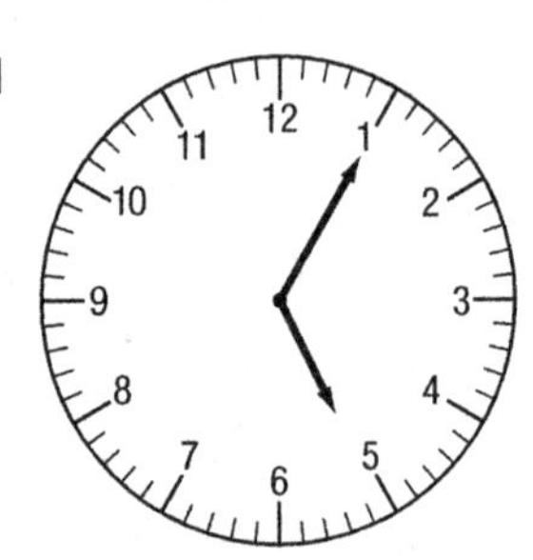

17 If each trip takes 26 minutes, calculate the arrival time for each boat if they depart at the following times.

a	2:30 p.m.	**b**	9:25 a.m.	**c**	11:45 a.m.	**d**	1:06 p.m.	**e**	4:47 p.m.
f	10:36 a.m.	**g**	5:17 p.m.	**h**	12:52 p.m.	**i**	8:28 a.m.	**j**	3:56 p.m.

18 Use these starting and finishing times to calculate the duration of each plane trip.

	Start	*Finish*		*Start*	*Finish*
a	8:40 a.m.	10:20 a.m.	**b**	2:55 p.m.	3:45 p.m.
c	11:30 a.m.	1:10 p.m.	**d**	10:05 a.m.	12:30 p.m.
e	4:15 p.m.	7:50 p.m.	**f**	9:10 a.m.	12:25 p.m.
g	3:40 p.m.	8:05 p.m.	**h**	2:40 p.m.	4:15 p.m.
i	1045	1320	**j**	1750	2035
k	0627	0814	**l**	0913	1107
m	1526	1657	**n**	1447	1725
o	1942	2144	**p**	2118	2339

19 Put each set of times in order from earliest to latest.

a	6:45 p.m. 1850 6:45 a.m.	**b**	2230 2145 10 p.m.
c	9:20 a.m. 0910 9:35 a.m.	**d**	1340 1:35 p.m. 1300
e	1705 4:55 p.m. 1640	**f**	8:15 a.m. 0800 2022
g	1428 2:30 p.m. 1400	**h**	1846 6 p.m. 0640
i	8 p.m. 0830 2005	**j**	10 a.m. 10 p.m. 1015
k	7:25 a.m. 1920 0700	**l**	1535 3:15 p.m. 1530
m	2015 8:15 a.m. 8 p.m.	**n**	0400 4:50 a.m. 1600

Strand Chance and Data

6.4.1 Collect and interpret locally relevant statistical data

Find the mean, median, mode and range

Remember

The **mean** is found by adding a set of scores together and dividing by the number of scores.
The **median** is the middle score or number of a set of scores or numbers. It is found by putting the numbers in order.
The **mode** is the score or number that occurs most often in a set of scores or numbers.
The **range** of a set of scores or numbers is the difference between the highest and lowest score or number.
For example, if the highest number is 20 and the lowest is 14, the range is 6 (20 – 14).

1 Find the mean of each set of numbers.

a 14, 21, 9, 100, 11
b 9, 6, 67, 13, 40
c 80, 95, 78, 15, 12
d 34, 17, 28, 67, 19, 21
e 46, 21, 19, 32, 54, 44
f 83, 29, 50, 45, 67, 62
g 75, 37, 54, 18, 25, 73
h 91, 66, 58, 85, 39, 87
i 49, 14, 73, 68, 64, 56
j 87, 77, 94, 81, 97, 38, 72
k 19, 26, 44, 12, 33, 47, 29
l 83, 96, 89, 71, 67, 95, 101
m 43, 39, 32, 49, 53, 45, 33
n 23, 18, 31, 16, 35, 47, 28, 42
o 87, 75, 95, 83, 98, 61, 77, 80

2 Find the median of each set of numbers.

a 21, 7, 43, 18, 25
b 36, 72, 31, 27, 16
c 8, 33, 14, 25, 17
d 95, 83, 98, 71, 84
e 53, 48, 60, 43, 57
f 23, 45, 19, 7, 102
g 76, 83, 91, 47, 88, 94, 80
h 65, 60, 58, 73, 69, 70, 54
i 109, 96, 54, 123, 87, 117, 92
j 158, 141, 127, 98, 124, 153, 146
k 213, 245, 189, 175, 208, 191, 228
l 74, 33, 65, 108, 57, 112, 143
m 34, 21, 17, 38, 46, 29, 33, 41, 55
n 48, 57, 39, 45, 62, 56, 81, 68, 73

3 Find the mode of each set of numbers.

a 3, 7, 4, 4, 9, 11, 4, 2, 4, 5, 4, 9
b 6, 12, 3, 5, 6, 8, 6, 4, 6, 12, 7, 2
c 11, 9, 13, 11, 6, 3, 11, 3, 7, 5, 11, 3
d 8, 8, 3, 6, 1, 3, 8, 6, 9, 8, 9, 8
e 32, 45, 31, 45, 32, 32, 33, 31, 32, 29
f 17, 19, 13, 15, 13, 12, 13, 17, 15, 12
g 22, 29, 23, 22, 23, 24, 25, 22, 25, 27
h 54, 57, 51, 54, 58, 51, 56, 51, 57, 55

4 Find the range of each set of numbers.

a 35, 78, 37, 64, 49, 53, 81, 76, 58, 60
b 17, 21, 9, 36, 28, 19, 11, 34, 41, 29
c 185, 211, 176, 148, 224, 192, 163, 171
d 435, 412, 456, 423, 397, 418, 405, 399
e 108, 97, 110, 89, 93, 105, 116, 98, 113
f 74, 68, 93, 81, 102, 79, 85, 92, 89

6.4.6 Estimate quantities and numbers

Estimate to find the correct answer

1 Use estimation to help you choose the correct answer for these multiplications.

a	71 × 39 =	2769	2139	3459	b	52 × 15 =	520	780	1020
c	12 × 99 =	1218	1188	948	d	25 × 17 =	175	300	425
e	30 × 48 =	840	1440	1120	f	48 × 24 =	1152	2502	832
g	37 × 41 =	1230	2357	1517	h	24 × 25 =	250	920	600
i	85 × 22 =	1870	1540	2010	j	96 × 13 =	1248	968	1608

2 Use estimation to help you choose the correct answer for these divisions.

a	768 ÷ 8 =	106	36	96	b	531 ÷ 9 =	59	89	19
c	861 ÷ 7 =	93	153	123	d	1734 ÷ 6 =	359	289	159
e	720 ÷ 48 =	20	15	50	f	984 ÷ 24 =	41	71	401
g	2142 ÷ 51 =	32	42	52	h	3016 ÷ 29 =	14	104	54
i	1568 ÷ 32 =	29	509	49	j	3708 ÷ 18 =	206	26	106

3 Use estimation to help you choose the correct answer for these decimal multiplications.

a	5.9 × 3 =	15.1	17.7	20	b	7.4 × 6 =	40	48.5	44.4
c	6.9 × 4 =	27.6	16.3	29.6	d	5.8 × 5 =	32	25	29
e	4.6 × 6 =	24.6	27.6	30.6	f	3.9 × 7 =	21.3	27.3	31.6
g	7.2 × 8 =	57.6	64.2	56.3	h	8.3 × 9 =	70.7	86.5	74.7
i	6.7 × 5 =	30.8	33.5	40.8	j	4.4 × 8 =	35.2	31.8	40.6
k	7.7 × 5 =	35	38.5	42.5	l	9.3 × 9 =	83.7	79.3	95.4
m	8.8 × 6 =	55.6	52.8	41.7	n	5.4 × 9 =	44.3	55.6	48.6
o	3.8 × 5 =	20.4	19	16.5	p	20.6 × 7 =	144.2	180	120.3
q	15.8 × 3 =	47.4	45.6	59.8	r	25.5 × 7 =	178.5	210.5	125.5
s	30.8 × 6 =	179.8	205.9	184.8	t	19.7 × 5 =	106.3	98.5	86.5
u	19.4 × 8 =	115.2	155.2	175.2	v	12.9 × 4 =	47.6	51.6	55.3

4 Try choosing the correct answers for these harder decimal multiplications.

a	11.38 × 5 =	54.9	56.9	63.7	b	10.67 × 5 =	50.15	58.75	53.35
c	12.63 × 3 =	35.19	37.89	40.19	d	15.97 × 4 =	82.18	63.88	56.78
e	25.14 × 8 =	201.12	196.12	98.2	f	11.69 × 8 =	98.92	93.52	81.42
g	20.05 × 6 =	160.5	110.3	120.3	h	50.56 × 4 =	194.4	202.24	310.4
i	95.89 × 5 =	479.45	501.45	511.5	j	21.75 × 7 =	136.5	140.75	152.25
k	60.91 × 5 =	350.15	304.55	405.5	l	24.83 × 4 =	102.2	86.52	99.32
m	19.66 × 6 =	125.16	117.96	96.76	n	49.37 × 3 =	148.11	167.51	131.61
o	98.76 × 5 =	493.8	520.8	412.8	p	30.23 × 7 =	250.31	211.61	196.81
q	23.87 × 6 =	113.2	143.22	181.2	r	35.95 × 4 =	143.8	121.8	167.8

6.4.7 Round off amounts

Round off whole numbers

1 Round these numbers to the nearest hundred.

a	2759	b	7312	c	5625	d	8856	e	4320
f	1678	g	3079	h	6982	i	9234	j	7880
k	16 759	l	67 812	m	35 091	n	43 537	o	26 708
p	83 145	q	58 662	r	10 179	s	92 250	t	76 520
u	576 412	v	813 937	w	498 367	x	717 250	y	206 892

2 Round these numbers to the nearest thousand.

a	3127	b	6786	c	4008	d	7950	e	4500
f	58 743	g	33 026	h	18 545	i	72 788	j	42 450
k	11 232	l	91 067	m	52 198	n	26 674	o	83 609
p	218 445	q	407 654	r	678 038	s	931 693	t	125 344
u	376 569	v	812 241	w	556 809	x	116 285	y	934 512

3 Copy and complete this chart by rounding each number to the nearest thousand and ten thousand.

	Number	Nearest 1000	Nearest 10 000
a	415 238		
b	630 744		
c	296 301		
d	154 876		
e	5 839 675		
f	3 267 148		
g	1 841 602		
h	7 255 813		
i	9 112 346		
j	8 370 617		
k	2 455 278		
l	6 087 502		
m	5 163 824		
n	4 159 087		
o	8 370 572		
p	1 402 663		

4 Copy and complete this chart by filling each space with a number that can be rounded as shown.

	Number	Nearest 1000	Nearest 10 000
a		675 000	680 000
b		467 000	470 000
c		216 000	220 000
d		787 000	790 000
e		323 000	320 000
f		912 000	910 000
g		155 000	150 000
h		847 000	850 000
i		259 000	260 000
j		138 000	140 000
k		595 000	600 000
l		414 000	410 000
m		386 000	390 000
n		623 000	620 000
o		706 000	710 000
p		271 000	270 000

Round off decimals

1 Round these decimal numbers to the nearest whole number.

a	23.8	**b**	54.3	**c**	17.9	**d**	60.5	**e**	8.2	**f**	41.7
g	12.87	**h**	36.12	**i**	65.05	**j**	54.75	**k**	27.18	**l**	78.35
m	37.09	**n**	93.27	**o**	86.19	**p**	40.23	**q**	11.66	**r**	98.55

2 Round these decimal numbers to the nearest tenth (one decimal place).

a	13.81	**b**	27.38	**c**	92.15	**d**	45.78	**e**	34.09	**f**	6.24
g	56.07	**h**	82.55	**i**	79.16	**j**	29.19	**k**	63.67	**l**	50.53
m	15.41	**n**	93.75	**o**	38.74	**p**	81.06	**q**	32.11	**r**	64.22

3 Round these decimal numbers to the nearest hundredth (two decimal places).

a	56.513	**b**	12.852	**c**	47.179	**d**	63.018	**e**	70.165	**f**	82.941
g	15.444	**h**	23.816	**i**	95.067	**j**	24.642	**k**	7.899	**l**	38.196
m	10.353	**n**	82.845	**o**	26.008	**p**	49.811	**q**	53.098	**r**	76.222

4 Write the decimal numbers that could be rounded to 23.

23.25 23.559 22.4 22.816 22.609
22.34 23.395 22.7 22.09 23.056

5 Write the decimal numbers that could be rounded to 18.

18.96 17.09 17.67 17.009 18.39
18.705 18.47 18.135 18.2 17.51

6 Write the decimal numbers that could be rounded to 35.6

35.63 36.25 35.58 35.62 35.61
35.617 36.71 35.635 35.59 35.578

7 Copy and complete this chart by rounding these decimal numbers to the nearest tenth and the nearest hundredth.

	Number	Nearest $\frac{1}{10}$	Nearest $\frac{1}{100}$
a	8.752		
b	5.089		
c	9.616		
d	8.075		
e	6.344		

	Number	Nearest $\frac{1}{10}$	Nearest $\frac{1}{100}$
f	7.953		
g	4.126		
h	9.583		
i	5.208		
j	3.455		

Round off money

1 Round these amounts of money to the nearest 10 toea.

a	K10.59	b	K15.25	c	K6.73	d	K11.56	e	K16.78	f	K8.06
g	K12.94	h	K9.86	i	K8.71	j	K5.45	k	K19.27	l	K17.96
m	K9.23	n	K21.47	o	K25.61	p	K14.85	q	K5.04	r	K7.62
s	K19.95	t	K16.32	u	K13.09	v	K22.18	w	K15.77	x	K23.14

2 Round these amounts of money to the nearest kina.

a	K23.45	b	K67.12	c	K75.15	d	K40.95	e	K29.90	f	K21.50
g	K17.81	h	K35.08	i	K63.35	j	K89.40	k	K56.10	l	K9.37
m	K49.45	n	K51.25	o	K73.05	p	K14.20	q	K28.67	r	K95.79
s	K81.66	t	K26.49	u	K7.83	v	K12.32	w	K85.27	x	K57.18

3 Round these fractions of toea to the nearest whole toea.

a	50.6t	b	23.8t	c	70.5t	d	15.2t	e	34.9t	f	78.9t
g	45.5t	h	67.1t	i	36.6t	j	53.4t	k	9.8t	l	91.3t
m	32.16t	n	49.35t	o	88.59t	p	57.13t	q	85.64t	r	72.66t
s	19.06t	t	31.99t	u	60.08t	v	42.58t	w	26.51t	x	43.27t

4 Which of these amounts could be rounded to K25?

K25.87 K24.50 K25.15 K25.95 K24.85
K24.10 K25.50 K24.65 K26.10 K25.37

5 Which of these amounts could be rounded to K68.50?

K68.57 K68.49 K68.45 K68.52 K68.75
K68.55 K68.43 K68.51 K68.47 K68.40

6 Which of these amounts could be rounded to K112.90?

K112.87 K112.95 K112.99 K112.85 K112.92
K112.83 K112.80 K112.88 K112.75 K112.96

7 Round these amounts of money to the nearest 10 kina.

a	K56.25	b	K254.99	c	K128.25	d	K175.05	e	K97.75
f	K714.50	g	K546.30	h	K71.04	i	K393.40	j	K106.72
k	K408.35	l	K298.79	m	K357.86	n	K811.10	o	K34.60
p	K175.45	q	K961.85	r	K899.05	s	K523.90	t	K275.40
u	K650.95	v	K481.24	w	K69.85	x	K337.12	y	K496.20

Round off temperature, time, capacity and weight

1 Round these temperatures to the nearest degree.

a	25.7°	**b**	24.2°	**c**	27.8°	**d**	21.6°	**e**	30.5°	**f**	29.1°
g	33.6°	**h**	22.3°	**i**	37.9°	**j**	34.5°	**k**	17.4°	**l**	20.7°
m	39.9°	**n**	31.6°	**o**	19.8°	**p**	23.3°	**q**	26.2°	**r**	38.5°

2 Round these times to the nearest second (s).

a	14.6 s	**b**	11.8 s	**c**	25.2 s	**d**	23.9 s	**e**	13.4 s	**f**	20.5 s
g	9.08 s	**h**	22.56 s	**i**	13.35 s	**j**	7.11 s	**k**	15.87 s	**l**	29.54 s
m	12.37 s	**n**	10.05 s	**o**	21.68 s	**p**	18.44 s	**q**	23.95 s	**r**	8.17 s

3 Round these times to the nearest tenth of a second (s).

a	11.83 s	**b**	15.06 s	**c**	12.57 s	**d**	9.48 s	**e**	15.94 s	**f**	13.35 s
g	17.59 s	**h**	16.14 s	**i**	14.21 s	**j**	8.03 s	**k**	10.66 s	**l**	11.78 s
m	16.92 s	**n**	12.65 s	**o**	18.13 s	**p**	23.54 s	**q**	21.78 s	**r**	19.36 s
s	14.67 s	**t**	10.24 s	**u**	20.51 s	**v**	24.17 s	**w**	17.88 s	**x**	15.97 s

4 Round these capacities to the nearest litre (L).

a	8.7 L	**b**	17.5 L	**c**	10.3 L	**d**	5.1 L	**e**	4.25 L	**f**	9.76 L
g	15.04 L	**h**	22.67 L	**i**	3.125 L	**j**	7.835 L	**k**	11.549 L	**l**	6.013 L
m	9.003 L	**n**	0.815 L	**o**	4.708 L	**p**	1.255 L	**q**	8.378 L	**r**	2.504 L
s	1078 mL	**t**	3450 mL	**u**	978 mL	**v**	5009 mL	**w**	2578 mL	**x**	7189 mL

5 Round these capacities to the nearest tenth of a litre (L).

a	5.37 L	**b**	2.18 L	**c**	3.05 L	**d**	0.82 L	**e**	4.51 L	**f**	2.93 L
g	1.24 L	**h**	7.89 L	**i**	5.06 L	**j**	1.77 L	**k**	3.52 L	**l**	0.35 L
m	2.653 L	**n**	8.147 L	**o**	1.565 L	**p**	4.038 L	**q**	5.789 L	**r**	3.107 L
s	0.845 L	**t**	2.722 L	**u**	6.343 L	**v**	9.116 L	**w**	1.007 L	**x**	4.888 L

6 Round these weights to the nearest kilogram (kg).

a	5.6 kg	**b**	3.8 kg	**c**	2.3 kg	**d**	1.47 kg	**e**	5.63 kg	**f**	0.89 kg
g	8.02 kg	**h**	6.35 kg	**i**	4.11 kg	**j**	5.27 kg	**k**	2.94 kg	**l**	4.89 kg
m	1.675 kg	**n**	2.417 kg	**o**	4.275 kg	**p**	9.081 kg	**q**	3.403 kg	**r**	0.617 kg
s	6589 g	**t**	2008 g	**u**	1831 g	**v**	9600 g	**w**	7105 g	**x**	3506 g

7 Round these weights to the nearest tenth of a tonne (t).

a	2.46 t	**b**	4.83 t	**c**	0.17 t	**d**	1.74 t	**e**	5.69 t	**f**	2.07 t
g	1.52 t	**h**	3.75 t	**i**	7.96 t	**j**	9.21 t	**k**	4.14 t	**l**	0.55 t
m	5.575 t	**n**	1.063 t	**o**	2.408 t	**p**	0.716 t	**q**	3.502 t	**r**	6.733 t

Assessment Chance and Data

Multiple choice test

1 What is the mean of 18, 13, 25, 11 and 33?

a 100 **b** 5 **c** 20 **d** 25

2 What is the mean of 43, 51, 72, 36, 47, 89, 31 and 55?

a 53 **b** 426 **c** 49 **d** 89

3 What is the median of 64, 53, 21, 48 and 59?

a 53 **b** 21 **c** 64 **d** 59

4 What is the median of 102, 87, 115, 96, 88, 125 and 109?

a 722 **b** 125 **c** 87 **d** 102

5 What is the mode of 37, 41, 28, 40, 37, 50, 41, 41, 45 and 40?

a 40 **b** 41 **c** 37 **d** 28

6 What is the range of 143, 212, 108, 164, 129, 207, 230 and 186?

a 108 **b** 230 **c** 122 **d** 200

7 Using estimation, the answer to 51 × 19 is:

a 900 **b** 1000 **c** 960 **d** 969

8 Using estimation, the answer to 1260 ÷ 63 is:

a 10 **b** 20 **c** 200 **d** 100

9 Using estimation, the answer to 5.7 × 6 is:

a 34.2 **b** 3.4 **c** 342 **d** 30

10 6157 rounded to the nearest hundred is:

a 6000 **b** 6150 **c** 6100 **d** 6200

11 15 519 rounded to the nearest thousand is:

a 16 000 **b** 15 000 **c** 15 500 **d** 15 520

12 623 450 rounded to the nearest ten thousand is:

a 624 000 **b** 62 000 **c** 620 000 **d** 630 000

13 39.4 rounded to the nearest whole number is:

a 40 **b** 39 **c** 394 **d** 390

14 76.25 rounded to the nearest tenth:

a 76.3 **b** 77 **c** 76.5 **d** 76.2

15 19.871 rounded to the nearest hundredth is:

a 20 **b** 19.9 **c** 19.88 **d** 19.87

16 63.055 rounded to the nearest tenth is:

a 63.06 **b** 63.1 **c** 63 **d** 63.05

17 Which of these amounts could be rounded to K32?

a K32.75 **b** K32.50 **c** K32.30 **d** K31.25

18 Which of these temperatures could be rounded to 25°?

a 24.3° **b** 25.2° **c** 25.7° **d** 24.1°

19 4547 mL rounded to the nearest litre is:

a 4 L **b** 45 L **c** 46 L **d** 5 L

Strand Patterns and Algebra

6.5.2 Explore number patterns

Identify and continue number patterns

1 Copy and complete each number sequence.

a 16, 20, 19, 23, 22, ____, ____, ____
b 30, 29, 27, 24, ____, ____, ____
c 7, 8, 10, 13, ____, ____, ____, ____
d 100, 80, 90, 70, 80, ____, ____, ____
e 2, 4, 8, 16, ____, ____, ____, ____
f 3, 7, 15, 31, ____, ____, ____, ____
g 50, 45, 44, 39, 38, ____, ____, ____
h 2, 7, 9, 14, 16, ____, ____, ____
i 21, 20, 17, 16, 13, ____, ____, ____
j 4, 8, 9, 13, 14, ____, ____, ____
k 8, 18, 17, 27, 26, ____, ____, ____
l 6, 9, 11, 14, 16, ____, ____, ____
m 1, 3, 6, 8, 11, 13, ____, ____, ____
n 5, 10, 9, 18, 17, 34, ____, ____, ____
o 3, 6, 12, 24, ____, ____, ____, ____
p 5, 10, 12, 24, 26, 52, ____, ____, ____
q 3, 9, 27, 81, ____, ____, ____
r 100, 90, 89, 79, 78, 68, ____, ____, ____
s 3, 6, 10, 15, 21, ____, ____, ____
t 1, 4, 9, 16, 25, ____, ____, ____, ____

2 Describe how each number sequence has been created. For example: 3, 9, 27, 81, 243, 729 has been created by multiplying by 3.

a 5, 11, 17, 23, 29, 35, 41
b 72, 65, 58, 51, 44, 37, 30
c 50, 40, 35, 25, 20, 10, 5
d 15, 16, 18, 21, 25, 30, 36
e 5, 10, 20, 40, 80, 160, 320
f 1, 5, 25, 125, 625, 3125
g 800, 400, 200, 100, 50, 25
h 3, 8, 7, 12, 11, 16, 15, 20, 19
i 2, 4, 3, 6, 5, 10, 9, 18, 17, 34
j 30, 29, 27, 24, 20, 15, 9, 2
k 1, 1, 2, 3, 5, 8, 13, 21, 34
l 2, 10, 9, 17, 16, 24, 23, 31, 30
m 2, 10, 11, 55, 56, 280, 281
n 50, 41, 33, 26, 20, 15, 11, 8, 6, 5
o 2, 11, 10, 19, 18, 27, 26, 35
p 2, 4, 5, 10, 11, 22, 23, 46, 47
q 81, 71, 68, 58, 55, 45, 42, 32
r 67, 65, 60, 58, 53, 51, 46, 44
s 3, 4.5, 6.5, 8, 10, 11.5, 13.5, 15
t 40, 38.75, 37.25, 36, 34.5, 33.25, 31.75

3 Create your own number sequences to fit each of these descriptions. You may start at a number of your choice but each sequence must have at least eight numbers.

a Add 4
b Subtract 3
c Multiply by 2
d Halve
e Add 12 and then subtract 1
f Double and then add 1
g Subtract 5 and then subtract 2
h Add 8 and then add 3
i Multiply by 4
j Divide by 10
k Subtract 4 and then add 1
l Add 12 and then subtract 3
m Add 2, then 4, then 6 and so on
n Subtract 1, then 2, then 3 and so on
o Add 5 and then subtract 1
p Double and then add 2
q Subtract 2.5 and then subtract 1.5
r Add 7 and then add 0.5

Apply given rules to complete tables

Apply the rule above each table to complete the IN and OUT columns.

1 + 4

IN	OUT
11	
16	
	31
33	
59	
	92

2 − 7

IN	OUT
22	
26	
35	
	33
	51
	76

3 × 3

IN	OUT
4	
9	
12	
20	
	75
	90

4 ÷ 5

IN	OUT
25	
	8
	10
100	
	50
325	

5 × 8

IN	OUT
6	
9	
	88
	120
21	
	320

6 + 15

IN	OUT
12	
	38
36	
	62
69	
	101

7 ÷ 6

IN	OUT
18	
30	
	9
	11
90	
108	

8 − 12

IN	OUT
21	
40	
	53
73	
96	
	115

9 × 2 and + 3

IN	OUT
5	
8	
12	
15	
30	
	93

10 − 5 and × 3

IN	OUT
10	
25	
28	
	75
	120
67	

11 ÷ 2 and + 1

IN	OUT
16	
28	
42	
	36
	47
114	

12 + 20 and ÷ 3

IN	OUT
7	
13	
	14
	17
37	
	22

13 − 3 and × 10

IN	OUT
	80
17	
	290
	440
61	
	820

14 ÷ 2 and − 5

IN	OUT
20	
	18
58	
	31
106	
	145

15 × 6 and + 8

IN	OUT
6	
11	
	98
	146
32	
	248

Identify rules and complete tables

Work out the rule that connects the numbers in each table and use it to complete each table.

1

IN	OUT
16	32
9	18
11	22
14	
30	
25	

2

IN	OUT
6	15
8	19
5	13
9	
15	
10	

3

IN	OUT
3	5
10	19
20	39
8	
5	
11	

4

IN	OUT
4	17
9	37
11	45
8	
2	
12	

5

IN	OUT
20	11
16	9
30	16
12	
10	
24	

6

IN	OUT
16	10
20	12
40	22
12	
50	
18	

7

IN	OUT
3	8
12	35
5	14
20	
11	
9	

8

IN	OUT
5	40
12	110
50	490
2	
17	
23	

9

IN	OUT
5	35
7	45
12	70
8	
11	
20	

10

IN	OUT
9	20
25	52
30	62
100	
21	
16	

11

IN	OUT
16	5
20	6
40	11
48	
100	
4	

12

IN	OUT
2	11
5	26
10	51
11	
20	
25	

13

IN	OUT
3	19
5	25
8	34
11	
15	
30	

14

IN	OUT
2	5
4	17
10	101
7	
5	
12	

15

IN	OUT
10	25
9	22
3	4
20	
24	
40	

6.5.3 Introduce pronumerals

Write and interpret pronumerals in formulae

Remember

A pronumeral is a letter or symbol that stands for an unknown value. For example: in the formula $a + 5 = 12$, a stands for 7.

Algebra does not use the multiplication sign. For example: instead of $a = b \times 4$, we write $a = 4b$. The number is always written first.

Use brackets when necessary to show the part of the formula that is to be calculated first.

1 Write each of these rules as a formula. For example: '*d* is the same as 4 added to *f*' becomes $d = f + 4$.

a x is the same as 5 added to y

b m is the same as 10 subtracted from n

c a is the same as 3 multiplied by b

d f is the same as g divided by 5

e s is the same as 8 added to t

f k is the same as 4 subtracted from p

g w is the same as x multiplied by 5

h c is the same as 12 divided by d

i b is the same as g minus 2

j t is the same as f plus 4 and 3

k m is the same as 4 times n

l y is the same as x divided by 7

2 Write each of these formulae as a rule in the same way as the previous activity.

a $d = x + 6$ **b** $c = g - 8$ **c** $m = 11n$ **d** $k = 3s$

e $y = 20 \div x$ **f** $a = b + 6 + 7$ **g** $x = y - 12$ **h** $f = g \div 2$

i $p = 5 + q$ **j** $s = 5t$ **k** $b = 3 + a + 7$ **l** $w = z - 8 - 3$

3 Write each of these rules as a formula. For example: 'to get *x*, first add 3 to *y* and then multiply by 6' becomes $x = 6(y + 3)$.

a To get m, first add 7 to n and then multiply by 5.

b To get p, first subtract 8 from q and then multiply by 10.

c To get a, first multiply b by 9 and then add 4.

d To get d, first divide c by 2 and then subtract 6.

e To get w, first multiply z by 7 and then subtract 3.

f To get g, first add 6 to h and then multiply by 12.

g To get x, first divide y by 4 and then add 10.

h To get k, first subtract 6 from j and then divide by 4.

i To get s, first add 15 to t and then multiply by 3.

j To get f, first divide g by 8 and then subtract 11.

k To get b, first multiply a by 6 and then add 7.

l To get y, first subtract 12 from x and then multiply by 5.

4 Write each of these formulae as a rule in the same way as the previous activity.

a	$g = 4h - 6$	**b**	$x = 5(y + 4)$	**c**	$m = n \div 2 + 3$	**d**	$a = 6(b - 7)$
e	$d = (c - 4) \div 3$	**f**	$s = 2t + 9$	**g**	$k = 11(j + 8)$	**h**	$f = 8 + g - 5$
i	$w = 7z + 10$	**j**	$h = 3(k - 8)$	**k**	$p = (5 + q) \div 2$	**l**	$y = 8x + 9$

Find the value of pronumerals

1 Find the value of the pronumeral in each of these number sentences.

a	$5b = 15$	**b**	$7y = 28$	**c**	$6p = 30$	**d**	$8m = 56$
e	$x + 3 = 15$	**f**	$27 - p = 12$	**g**	$a + 17 = 31$	**h**	$d - 11 = 16$
i	$33 \div c = 3$	**j**	$g \div 9 = 7$	**k**	$240 \div f = 10$	**l**	$n \div 8 = 8$
m	$s - 17 = 22$	**n**	$31 + q = 58$	**o**	$43 - k = 28$	**p**	$t + 27 = 52$
q	$4f = 64$	**r**	$45 + z = 81$	**s**	$h - 29 = 54$	**t**	$108 \div p = 12$
u	$72 - r = 47$	**v**	$11n = 77$	**w**	$49 \div e = 7$	**x**	$g + 57 = 91$

2 What is the pronumeral worth in each of these longer number sentences?

a	$6 + y + 15 = 33$	**b**	$35 - 8 - a = 19$	**c**	$3n + 6 = 36$
d	$52 - g - 13 = 28$	**e**	$15 + 18 + x = 64$	**f**	$b - 9 - 12 = 7$
g	$5m - 50 = 100$	**h**	$k + 17 + 9 = 45$	**i**	$5 + 5w = 105$
j	$30 \div 6 + p = 14$	**k**	$7t + 35 = 70$	**l**	$44 - g - 8 = 22$
m	$17 + 15 - d = 23$	**n**	$z + 20 - 10 = 40$	**o**	$19 + y - 7 = 22$
p	$30 - 16 + f = 20$	**q**	$25 - a + 15 = 35$	**r**	$s - 12 - 15 = 21$
s	$35 \div 5 + k = 16$	**t**	$23 - 7d = 2$	**u**	$6p - 18 = 24$
v	$12 + 9 + 3h = 30$	**w**	$4r - 25 - 7 = 48$	**x**	$24 \div 3 - m = 5$

3 Find the value of the pronumeral in each of these number sentences.

a	$6 + 7 = x + 8$	**b**	$3 \times 5 = 6y + 3$	**c**	$12 - 5 = 4 + s$
d	$p - 7 = 5 + 9$	**e**	$11z - 10 = 50 - 5$	**f**	$36 \div 4 = 3a + 3$
g	$8 \times 12 = 4t - 4$	**h**	$42 \div b = 2 \times 3$	**i**	$23 + d = 46 - 17$
j	$30 \times 3 = w + 66$	**k**	$g - 15 = 35 \div 7$	**l**	$200 - f = 20 \times 7$
m	$72 \div 9 = 20 - h$	**n**	$3 \times 2 \times 5 = x - 10$	**o**	$17 + y + 6 = 3 \times 11$
p	$81 - 17 - 9 = 3n - 5$	**q**	$56 - m = 16 + 7 + 15$	**r**	$15 \times 4 = 10v + 10$
s	$60 \div 12 = 17 - k$	**t**	$13 + 18 = 4 + 3b$	**u**	$s - 7 = 5 \times 2 + 3$
v	$24 \div 8 = p - 6 - 9$	**w**	$6 \times 3 \times 10 = 18r$	**x**	$2u - 6 = 9 + 8 + 11$

Remember

Brackets show the part/s of a number sentence that is/are to be calculated first.

4 What is the value of the pronumeral in each of these number sentences?

a $7(x + 8) = 63$

b $9(n - 7) = 27$

c $(3 \times 6) - (100 \div a) = 8$

d $3(y + 6) = 24$

e $6(g - 2) = 48$

f $4(25 - b) = 80$

g $36 \div (m + 9) = 3$

h $(3 \times 7) + (p \times 3) = 33$

i $2(7 + p + 8) = 40$

j $(30 \div 5) - (d \div 4) = 2$

k $3(8 + h) - 8 = 22$

l $25 - (4 \times 5) + s = 17$

m $(w + 4) - (3 + 6) = 3$

n $9(9 + c) = 180$

o $37 - (d \div 2) = 28$

p $45 \div (z - 6) = 9$

q $(40 \div 8) + (7 \times f) = 26$

r $7(k + 9) - 10 = 74$

s $12(8 - x) = 72$

t $54 \div (y - 5) = 9$

u $(g \div 3) - (36 \div 6) = 4$

v $(a + 7) \div 2 = 10$

w $(m \div 5) - 3 = 7$

x $10(q - 3) + 8 = 48$

Assessment

1 Copy and complete each number sequence.

a 43, 42, 39, 38, 35, ____, ____, ____

b 8, 12, 11, 15, 14, ____, ____, ____

c 2, 4, 5, 10, 11, 22, ____, ____, ____

d 55, 54, 52, 51, 49, ____, ____, ____

e 30, 32.5, 34, 36.5, 38, 40.5, ____, ____, ____

f 26, 24, 23.5, 21.5, 21, ____, ____, ____

2 Describe how each number sequence has been created.

a 61, 55, 49, 43, 37, 31, 25, 19

b 18, 19, 24, 25, 30, 31, 36, 37

c 44, 46, 41, 43, 38, 40, 35, 37

d 2, 4, 6.5, 13, 15.5, 31, 33.5, 67

e 80, 79.5, 78.5, 78, 77, 76.5, 75.5, 75

f 1, 3, 4, 12, 13, 39, 40, 120, 121

3 Copy and complete each table.

a $\times 4$

IN	OUT
3	
8	
	40
12	
20	
50	

b $\div 3$

IN	OUT
6	
12	
30	
	11
	30
150	

c $\times 2$ and $+ 5$

IN	OUT
8	
15	
	45
24	
	69
40	

d $\div 2$ and $- 3$

IN	OUT
10	
18	
22	
	12
	22
60	

4 Work out the rule that connects the numbers in each table.

a

IN	OUT
3	15
5	25
8	40
11	55
20	100
25	125

b

IN	OUT
20	9
8	3
12	5
50	24
36	17
100	49

c

IN	OUT
5	12
7	16
50	102
10	22
9	20
24	50

d

IN	OUT
10	100
8	64
12	144
3	9
5	25
7	49

5 Write each rule as a formula.

a x is the same as 10 added to y

b f is the same as g multiplied by 4

c m is the same as 6 subtracted from n

d a is the same as b divided by 12

e s is the same as 5 times t

f w is the same as z minus 8

6 Find the value of the pronumeral in each number sentence.

a $7x = 42$

b $p + 8 = 17$

c $f \div 4 = 9$

d $y - 10 = 5$

e $3a + 6 = 27$

f $18 - 4b = 10$

g $16 - w - 7 = 4$

h $8 \times 3 = 30 - d$

i $36 \div g = 3 \times 3$

j $8 + 9 = 3k + 2$

k $9(h + 5) = 81$

l $(2 \times 8) + (c \times 4) = 36$

Important Facts

Length and Perimeter

10 millimetres (mm) = 1 centimetre (cm)

100 centimetres (cm) = 1 metre (m)

1000 metres (m) = 1 kilometre (km)

1 mm = $\frac{1}{10}$ cm = 0.1 cm

1 cm = $\frac{1}{100}$ m = 0.01 m

1 m = $\frac{1}{1000}$ km = 0.001 km

Time

60 seconds = 1 minute

60 minutes = 1 hour

24 hours = 1 day

7 days = 1 week

2 weeks = 1 fortnight

52 weeks = 1 year

12 months = 1 year

365 days = 1 year

366 days = 1 leap year

10 years = 1 decade

100 years = 1 century

1000 years = 1 millennium

a.m. = ante meridiem – the period between 12 midnight and 12 noon

p.m. = post meridiem – the period between 12 noon and 12 midnight

12-hour time divides a day into two 12-hour periods: a.m. and p.m.

24-hour time divides a day into twenty-four 1-hour periods, starting at midnight

Capacity

1000 millilitres (mL) = 1 litre (L)

1 mL = $\frac{1}{1000}$ L = 0.001 L

Weight

1000 grams (g) = 1 kilogram (kg)

1000 kilograms (kg) = 1 tonne (t)

1 g = $\frac{1}{1000}$ kg = 0.001 kg

1 kg = $\frac{1}{1000}$ t = 0.001 t

Money

100 toea (t) = 1 kina (K)

1t = $\frac{1}{100}$ K = K0.01

Area

10 000 square centimetres (cm^2) = 1 square metre (m^2)

10 000 square metres (m^2) = 1 hectare (ha)

100 hectares (ha) = 1 square kilometre (km^2)

Area of a rectangle = Length × Width

Area of a triangle = (Height × Base) ÷ 2

Temperature

Temperature is usually measured in degrees Celsius (°C). On the Celsius scale, water freezes at 0 degrees and boils at 100 degrees.

Volume

100 000 cubic centimetres (cm^3) = 1 cubic metre (m^3)

Volume of a rectangular prism = Length × Width × Height

Volume of a triangular prism = (Area of base) × Height

Angles

Angles are measured in degrees.

A right angle = 90°

A straight angle = 180°

A full turn = 360°

Acute angles measure between 0° and 90°.

Obtuse angles measure between 90° and 180°.

Reflex angles measure between 180° and 360°.

Place value

The value of a digit in a number is determined by its place value or position.

Millions	Hundreds of thousands	Tens of thousands	Thousands	Hundreds	Tens	Ones	Tenths	Hundredths	Thousandths

Square numbers

All square numbers can be arranged in a square pattern of dots.

For example:

• • • • $(16 = 4 \times 4 = 4^2)$
• • • •
• • • •
• • • •

$1^2 = 1$ $2^2 = 4$ $3^2 = 9$ $4^2 = 16$

$5^2 = 25$ $6^2 = 36$ $7^2 = 49$ $8^2 = 64$

$9^2 = 81$ $10^2 = 100$ $11^2 = 121$ $12^2 = 144$

Order of operations

Do **brackets** first, then **of**, then **multiplication** and **division** in the order they appear from left to right, and finally **addition** and **subtraction** in the order they appear from left to right.

Triangular numbers

All triangular numbers can be arranged in a triangular pattern of dots.

For example:

•
• •
• • •
• • • •

(10)

Triangular numbers from 1 to 100 are:
1, 3, 6, 10, 15, 21, 28, 36, 45, 55, 66, 78 and 91.

Answers

Strand: Number and Application

Pages 2–4

1 a 16369 b 12033 c 21209 d 13701
e 34341 f 56822 g 46700 h 81532
i 74312 j 135308 k 74549 l 48629

2 a 2731 b 2754 c 2342 d 23985
e 21872 f 24388 g 66494 h 41445
i 43783 j 13107 k 52361 l 28581

3 a 39774 b 30285 c 40752 d 19264
e 218912 f 61625 g 97224 h 397908
i 264945 j 742463 k 538092 l 1795136
m 586175 n 716220 o 1108351 p 809727

4 a 1243 b 3784 c 633 d 991
e 2589 r 3 f 6523 r 1 g 3795 r 1 h 3996 r 4
i 1935 r 9 j 1953 r 2 k 2139 r 12 l 874 r 22
m 946 r 18 n 1361 r 6 o 654 r 18 p 2947 r 2

5 a 33675, 43431, 14477, 5738 b 3584, 50176, 32477, 55551
c 44260, 11065, 99585, 61617 d 33646, 45633, 2173, 54325
e 70164, 32035, 6407, 66349 f 4506, 76895, 28918, 202426
g 38292, 12764, 76584, 125531 h 45955, 37008, 3084, 86352

6 a 31923 b 119831 c 12631 d 6592
e 16167 f 45508 g 4768 h 73604
i 3229 j 85514 k 24524 l 74671

7 a 24336 b 47339 c 59142 d 58806
e 29323 f 54170 g 27582 h 40895

8 a 87756 b 64427 c 78271 d 90300
e 67890 f 80361 g 82032 h 78773

9 a 3852 b 2153 c 5947 d 2981

10 a 24080 b 34108 c 48461 d 32202

Pages 4–5

1 a $\frac{1}{2}$ b $\frac{2}{4}, \frac{1}{2}$ c $\frac{1}{3}$ d $\frac{1}{5}$
e $\frac{3}{5}$ f $\frac{1}{4}$ g $\frac{1}{3}$ h $\frac{2}{5}$
i $\frac{2}{3}$ j $\frac{1}{4}$ k $\frac{5}{20}, \frac{1}{4}$ l $\frac{5}{50}, \frac{2}{20}, \frac{1}{10}$
m $\frac{1}{6}$ n $\frac{2}{8}, \frac{1}{4}$ o $\frac{5}{7}$ p $\frac{1}{3}$
q $\frac{2}{12}, \frac{1}{6}$ r $\frac{3}{7}$

2 There are many possible answers. Two are given for each question.
a $\frac{2}{8}, \frac{2}{12}$ b $\frac{4}{6}, \frac{6}{9}$ c $\frac{2}{12}, \frac{3}{18}$ d $\frac{8}{10}, \frac{12}{15}$
e $\frac{14}{20}, \frac{70}{100}$ f $\frac{2}{16}, \frac{3}{24}$ g $\frac{6}{14}, \frac{9}{21}$ h $\frac{2}{10}, \frac{3}{15}$
i $\frac{6}{8}, \frac{75}{100}$ j $\frac{10}{16}, \frac{15}{24}$ k $\frac{14}{18}, \frac{21}{27}$ l $\frac{4}{10}, \frac{6}{15}$
m $\frac{10}{14}, \frac{15}{21}$ n $\frac{6}{22}, \frac{9}{33}$ o $\frac{8}{18}, \frac{12}{27}$ p $\frac{10}{24}, \frac{15}{36}$
q $\frac{16}{18}, \frac{24}{27}$ r $\frac{10}{12}, \frac{15}{18}$

3 a $\frac{30}{40}$ b $\frac{10}{24}$ c $\frac{15}{24}$ d $\frac{2}{3}$ e $\frac{2}{5}$ f $\frac{8}{18}$
g $\frac{2}{5}$ h $\frac{8}{12}$ i $\frac{4}{5}$ j $\frac{42}{48}$

4 a $\frac{1}{2}$ b $\frac{5}{9}$ c $\frac{1}{3}$ d $\frac{1}{4}$ e $\frac{3}{4}$ f $\frac{3}{5}$
g $\frac{1}{5}$ h $\frac{2}{7}$ i $\frac{4}{5}$ j $\frac{2}{5}$ k $\frac{3}{4}$ l $\frac{1}{3}$
m $\frac{2}{5}$ n $\frac{3}{5}$ o $\frac{3}{4}$ p $\frac{2}{5}$ q $\frac{1}{7}$ r $\frac{1}{12}$

5 a $\frac{4}{12}, \frac{2}{6}, \frac{10}{30}$ b $\frac{25}{30}, \frac{15}{18}$ c $\frac{3}{4}, \frac{75}{100}, \frac{6}{8}$ d $\frac{4}{6}, \frac{2}{3}, \frac{8}{12}$
e $\frac{12}{32}, \frac{9}{24}, \frac{30}{80}$ f $\frac{2}{10}, \frac{4}{20}, \frac{3}{15}$ g $\frac{8}{10}, \frac{4}{5}$ h $\frac{10}{35}, \frac{6}{21}, \frac{14}{49}$
i $\frac{1}{6}, \frac{3}{18}$ j $\frac{18}{20}, \frac{90}{100}$

Pages 5–6

1 $\frac{15}{7}, \frac{10}{10}, \frac{9}{4}, \frac{11}{7}, \frac{7}{2}, \frac{8}{3}, \frac{5}{3}, \frac{10}{5}, \frac{8}{6}, \frac{19}{4}$

2 a $2\frac{1}{7}$ b $3\frac{1}{2}$ c $3\frac{1}{3}$ d $4\frac{3}{4}$ e $2\frac{6}{7}$ f $1\frac{1}{3}$
g $2\frac{3}{5}$ h $2\frac{5}{9}$ i $1\frac{2}{3}$ j $1\frac{3}{5}$ k $2\frac{3}{4}$ l $4\frac{2}{3}$
m $2\frac{2}{7}$ n $3\frac{1}{2}$ o $8\frac{1}{2}$ p $1\frac{6}{7}$ q $2\frac{1}{4}$ r $1\frac{2}{3}$

3 a $\frac{11}{2}$ b $\frac{23}{3}$ c $\frac{19}{5}$ d $\frac{15}{8}$ e $\frac{19}{4}$ f $\frac{17}{6}$
g $\frac{33}{4}$ h $\frac{29}{10}$ i $\frac{55}{9}$ j $\frac{38}{7}$ k $\frac{13}{9}$ l $\frac{29}{8}$
m $\frac{39}{4}$ n $\frac{25}{2}$ o $\frac{59}{8}$ p $\frac{33}{5}$ q $\frac{32}{7}$ r $\frac{107}{12}$

4 a $2\frac{5}{9}$ b $\frac{9}{4}$ c $\frac{15}{6}$ d $2\frac{4}{5}$ e $6\frac{2}{3}$ f $4\frac{7}{9}$
g $3\frac{2}{5}$ h $7\frac{3}{4}$ i $3\frac{3}{5}$ j $\frac{26}{7}$ k $1\frac{7}{8}$ l $3\frac{4}{5}$
m $\frac{25}{4}$ n $3\frac{1}{3}$ o $4\frac{5}{7}$ p $9\frac{2}{3}$ q $\frac{42}{5}$ r $7\frac{3}{4}$
s $\frac{25}{6}$ t $4\frac{1}{9}$ u $\frac{29}{8}$

Page 7

1 a $\frac{4}{9}$ b $1\frac{1}{3}$ c $\frac{6}{7}$ d $\frac{1}{4}$ e $1\frac{2}{3}$ f $\frac{5}{11}$
g $\frac{9}{20}$ h $1\frac{1}{10}$ i $\frac{2}{5}$ j $2\frac{1}{5}$ k $1\frac{3}{4}$ l $\frac{5}{12}$
m $1\frac{6}{7}$ n $\frac{1}{2}$ o $\frac{2}{3}$ p $1\frac{3}{4}$

2 a $\frac{41}{100}$ b $\frac{2}{5}$ c $\frac{3}{8}$ d $1\frac{2}{5}$ e $\frac{1}{6}$ f $1\frac{1}{12}$
g $1\frac{1}{3}$ h $\frac{8}{15}$ i $1\frac{1}{6}$ j $\frac{1}{12}$ k $\frac{1}{21}$ l $1\frac{13}{24}$
m $\frac{17}{40}$ n $1\frac{3}{20}$ o $\frac{13}{24}$ p $\frac{19}{60}$

3 EXCELLENT

Page 8

1 $5\frac{1}{3}$ **2** $8\frac{1}{2}$ **3** $5\frac{2}{5}$ **4** $6\frac{1}{8}$ **5** $5\frac{1}{4}$
6 $7\frac{7}{10}$ **7** $8\frac{1}{2}$ **8** $4\frac{3}{4}$ **9** $7\frac{3}{10}$ **10** $9\frac{3}{8}$
11 $3\frac{4}{9}$ **12** $9\frac{1}{12}$ **13** $7\frac{1}{12}$ **14** $6\frac{1}{15}$ **15** $4\frac{1}{24}$
16 $6\frac{5}{28}$ **17** $3\frac{19}{24}$ **18** $7\frac{31}{40}$

Page 9

1 $1\frac{1}{2}$ **2** $2\frac{4}{5}$ **3** $2\frac{1}{2}$ **4** $2\frac{9}{10}$ **5** $1\frac{5}{8}$
6 $2\frac{5}{6}$ **7** $\frac{7}{9}$ **8** $2\frac{3}{4}$ **9** $1\frac{3}{5}$ **10** $1\frac{5}{6}$
11 $2\frac{1}{10}$ **12** $2\frac{8}{21}$ **13** $\frac{8}{15}$ **14** $3\frac{11}{20}$ **15** $2\frac{23}{24}$
16 $2\frac{1}{18}$ **17** $1\frac{19}{84}$ **18** $1\frac{19}{30}$

Pages 10–11

1 **a** $1\frac{2}{3}$ **b** $1\frac{3}{5}$ **c** 3 **d** 2 **e** $1\frac{3}{4}$ **f** $3\frac{3}{5}$
g $2\frac{1}{2}$ **h** $2\frac{5}{8}$ **i** $2\frac{6}{7}$ **j** $3\frac{3}{5}$ **k** $1\frac{7}{9}$ **l** $1\frac{3}{7}$
m $\frac{3}{5}$ **n** $6\frac{2}{5}$ **o** $3\frac{3}{4}$

2 **a** 15 **b** 6 **c** 8 **d** 4 **e** 42 **f** 8
g 15 **h** 6 **i** 8 **j** 12 **k** 9 **l** 25
m 25 **n** 15 **o** 80

3 **a** $\frac{1}{10}$ **b** $\frac{1}{3}$ **c** $\frac{4}{15}$ **d** $\frac{1}{20}$ **e** $\frac{1}{3}$ **f** $\frac{3}{14}$
g $\frac{1}{32}$ **h** $\frac{4}{7}$ **i** $\frac{1}{8}$ **j** $\frac{9}{14}$ **k** $\frac{1}{3}$ **l** $\frac{5}{16}$
m $\frac{1}{6}$ **n** $\frac{6}{35}$ **o** $\frac{21}{40}$

4 **a** $1\frac{3}{4}$ **b** $2\frac{1}{3}$ **c** $2\frac{1}{16}$ **d** $3\frac{7}{16}$ **e** $3\frac{9}{25}$ **f** $9\frac{7}{12}$
g $2\frac{5}{8}$ **h** $10\frac{1}{30}$ **i** $8\frac{5}{16}$ **j** 6 **k** $2\frac{1}{4}$ **l** $8\frac{1}{3}$
m $3\frac{7}{36}$ **n** $4\frac{9}{28}$ **o** $8\frac{7}{16}$

5 **a** 12 **b** $\frac{15}{56}$ **c** 10 **d** $3\frac{1}{9}$ **e** $2\frac{13}{21}$ **f** 35
g $5\frac{5}{8}$ **h** $\frac{5}{14}$ **i** $10\frac{4}{5}$ **j** $\frac{27}{28}$ **k** 28 **l** $20\frac{4}{9}$

Pages 11–12

1 **a** 21 **b** $\frac{5}{18}$ **c** $6\frac{6}{7}$ **d** $\frac{1}{18}$ **e** $6\frac{2}{3}$ **f** $\frac{3}{14}$
g $\frac{1}{12}$ **h** $10\frac{2}{3}$ **i** $\frac{1}{14}$ **j** 15 **k** $\frac{5}{42}$ **l** $5\frac{2}{5}$
m 5 **n** $\frac{5}{64}$ **o** $5\frac{5}{8}$ **p** $\frac{1}{18}$ **q** $11\frac{3}{7}$ **r** $\frac{5}{84}$

2 **a** $\frac{5}{6}$ **b** $3\frac{3}{7}$ **c** $1\frac{1}{5}$ **d** $\frac{3}{7}$ **e** $1\frac{3}{5}$ **f** $3\frac{15}{16}$
g $1\frac{1}{6}$ **h** $1\frac{2}{3}$ **i** $\frac{2}{5}$ **j** $\frac{1}{4}$ **k** 2 **l** $2\frac{2}{7}$
m $1\frac{3}{25}$ **n** $2\frac{1}{7}$ **o** $\frac{16}{27}$ **p** $\frac{11}{12}$ **q** $2\frac{1}{3}$ **r** $\frac{5}{9}$

3 **a** $1\frac{9}{35}$ **b** $1\frac{7}{13}$ **c** $2\frac{1}{4}$ **d** $2\frac{2}{39}$ **e** $1\frac{67}{88}$ **f** $1\frac{17}{33}$
g $12\frac{1}{12}$ **h** $10\frac{2}{7}$ **i** $2\frac{1}{6}$ **j** $\frac{50}{69}$ **k** $2\frac{23}{36}$ **l** $1\frac{47}{161}$
m $2\frac{22}{25}$ **n** $1\frac{1}{18}$ **o** $1\frac{31}{60}$

4 **a** $2\frac{4}{13}, \frac{5}{8}, \frac{3}{32}, \frac{5}{6}, 14\frac{2}{7}$ **b** $\frac{9}{100}, 4\frac{12}{17}, 2\frac{1}{16}, 4\frac{4}{11}, 1\frac{5}{11}$
c $1\frac{67}{104}, 2\frac{5}{6}, 2\frac{4}{15}, 3\frac{3}{32}, 2\frac{5}{7}$

Pages 13–15

1 **a** 47.9 **b** 8.53 **c** 80.3 **d** 5.63
e 83.41 **f** 44.21 **g** 13.435 **h** 15.722
i 62.74 **j** 38.55 **k** 33.08 **l** 17.522
m 88.045 **n** 203.36 **o** 54.308 **p** 211.75

2 **a** 21.3 **b** 4.86 **c** 31.77 **d** 42.78
e 4.794 **f** 9.571 **g** 81.7 **h** 2.443
i 47.52 **j** 45.26 **k** 38.74 **l** 45.88
m 57.542 **n** 58.53 **o** 27.66 **p** 46.662

3 **a** 97.915 **b** 46.28 **c** 35.73 **d** 351.36
e 413.435 **f** 163.34 **g** 146.35 **h** 129.033

4 **a**

+	12.6	5.07	3.92	4.552	1.8
4.5	17.1	9.57	8.42	9.052	6.3
2.09	14.69	7.16	6.01	6.642	3.89
11.8	24.4	16.87	15.72	16.352	13.6
5.075	17.675	10.145	8.995	9.627	6.875
31.7	44.3	36.77	35.62	36.252	33.5

b

–	43.2	17.81	173.5	4.086	87.3
27.5	15.7	9.69	146	23.414	59.8
9.07	34.13	8.74	164.43	4.984	78.23
37.98	5.22	20.17	135.52	33.894	49.32
6.154	37.046	11.656	167.346	2.068	81.146
118.6	75.4	100.79	54.9	114.514	31.3

5 **a** 32.534 **b** 25.338 **c** 135.185 **d** 46.67
e 42.037 **f** 62.74 **g** 54.163

6 **a** 80.032 **b** 34.872 **c** 50.247 **d** 85.137

7 **a** 48.363 **b** 72.663 **c** 51.593 **d** 93.243

8 **a** 55.665 **b** 46.945 **c** 64.425 **d** 96.127

9 **a** 54.252 **b** 19.338 **c** 41.878 **d** 33.928

10 **a** 90.588 **b** 32.828 **c** 42.498 **d** 90.158

11 **a** 28.232 **b** 47.812 **c** 39.722 **d** 96.602

12 **a** 128.09 **b** 164.59 **c** 85.978 **d** 116.058

13 **a** 77.59 **b** 35.61 **c** 164.51 **d** 74.78

Pages 15–16

1 **a** 23.5 **b** 37.8 **c** 14.8 **d** 34.8
e 10.5 **f** 19.6 **g** 86.1 **h** 119.5
i 05.6 **j** 67.6 **k** 165.6 **l** 124.5
m 468 **n** 726.3 **o** 281.4 **p** 388.2
q 345.1 **r** 379

2 **a** K13.04 **b** K15.51 **c** K19.40 **d** K31.80
e K28.38 **f** K14.35 **g** K41.65 **h** K29.28
i K14.32 **j** K18.75 **k** K30.08 **l** K34.92
m K52.50 **n** K43.35 **o** K32.94 **p** K37.52

3 **a** 53.4 **b** 37.2 **c** 20.88 **d** 68.5
e 173.6 **f** 42.96 **g** 17.008 **h** 11.394
i 87.78 **j** 88.893 **k** 28.365 **l** 214.8
m 62.307 **n** 75.45 **o** 42.75 **p** 194.8
q 225.28 **r** 21.27 **s** 144.72 **t** 77.36
u 13.925 **v** 167.7 **w** 260.12 **x** 52.98

4

	Number of items and cost per item	Total cost	Change from K50
a	4 @ K6.70 each	K26.80	K23.20
b	3 @ K11.35 each	K34.05	K15.95
c	5 @ K7.50 each	K37.50	K12.50
d	10 @ K4.25 each	K42.50	K7.50
e	3 @ K9.60 each	K28.80	K21.20
f	6 @ K7.20 each	K43.20	K6.80
g	4 @ K3.95 each	K15.80	K34.20
h	8 @ K5.35 each	K42.80	K7.20
i	6 @ K6.55 each	K39.30	K10.70
j	5 @ K9.25 each	K46.25	K3.75

Pages 17–18

1 4.3 **2** 3.1 **3** 2.1 **4** 2.2 **5** 3.12
6 4.22 **7** 21.3 **8** 12.1 **9** 1.2 **10** 1.2
11 1.5 **12** 1.2 **13** 1.42 **14** 2.61 **15** 1.41
16 1.31 **17** 1.935 **18** 0.294 **19** 0.326 **20** 0.405
21 3.41 **22** 6.83 **23** 3.65 **24** 5.96 **25** 6.72
26 8.51 **27** 29.61 **28** 9.57 **29** 0.848 **30** 39.4
31 47.3 **32** 0.667

33

	Number of items and cost per item	Cost	Change from K100
a	8 @ K7.35 each	K58.80	K41.20
b	6 @ K11.65 each	K69.90	K30.10
c	3 @ K25.40 each	K76.20	K23.80
d	5 @ K16.95 each	K84.75	K15.25
e	2 @ K36.80 each	K73.60	K26.40
f	6 @ K12.80 each	K76.80	K23.20
g	5 @ K13.50 each	K67.50	K32.50
h	3 @ K31.25 each	K93.75	K6.25
i	7 @ K5.65 each	K39.55	K60.45
j	8 @ K8.60 each	K68.80	K31.20

Pages 19–20

1 a K47.95 b K27.40 c K61.65 d K34.25
2 a 24.85 km b 142.55 km
3 K78.65 **4** K59.85 **5** 466.5 L
6 K76.88 **7** K760.05 **8** K146.35
9 a 14.75 kg b 9.05 kg c 194 kg
10 a 4.565 kg b 24.265 kg
11 a 289.6 km b 463.36 km c 347.52 km d 521.28 km
12 5.605 L **13** 2.36 m **14** 13.4 kg **15** K1046.30

Pages 20–21

1 a 0.3 b 0.27 c 0.8 d 0.51 e 0.73 f 0.1
g 0.44 h 0.5 i 0.16 j 0.81 k 0.9 l 0.39
m 0.09 n 0.05 o 0.3 p 0.9 q 0.02 r 0.1
s 0.6 t 0.72 u 0.65 v 0.07 w 0.89 x 0.2

2 a 0.4 b 0.52 c 0.25 d 0.75 e 0.18 f 0.5
g 0.05 h 0.06 i 0.6 j 0.3 k 0.44 l 0.2
m 0.88 n 0.75 o 0.54 p 0.04 q 0.04 r 0.65
s 0.15 t 0.25 u 0.24 v 0.4 w 0.4 x 0.6

3 a 1.75 b 2.5 c 1.7 d 3.11 e 5.1 f 2.25
g 7.29 h 4.4 i 8.12 j 1.2 k 2.9 l 3.5
m 5.2 n 3.44 o 7.05 p 2.3 q 4.84 r 1.06
s 2.6 t 9.44 u 3.36 v 1.45 w 5.18 x 4.05

4 a 2.2 b 3.24 c 1.7 d 2.05 e 1.83 f 5.9
g 4.2 h 3.25 i 4.5 j 1.4 k 6.75 l 2.34
m 3.24 n 5.65 o 7.8 p 2.28 q 1.55 r 9.75
s 12.5 t 8.6 u 5.12 v 6.25 w 1.06 x 12.8

5 a $\frac{7}{10}$ b $\frac{63}{100}$ c $\frac{1}{5}$ d $\frac{87}{100}$ e $\frac{3}{5}$ f $\frac{31}{100}$
g $1\frac{1}{2}$ h $2\frac{3}{10}$ i $5\frac{4}{5}$ j $3\frac{2}{5}$ k $1\frac{1}{4}$ l $2\frac{3}{4}$
m $3\frac{3}{25}$ n $1\frac{11}{25}$ o $\frac{17}{20}$ p $8\frac{9}{25}$ q $5\frac{2}{25}$ r $7\frac{27}{50}$
s $2\frac{1}{25}$ t $19\frac{3}{5}$ u $1\frac{79}{100}$ v $3\frac{3}{20}$ w $11\frac{4}{5}$ x $10\frac{1}{20}$

6 a 0.43, $\frac{7}{10}$, $\frac{93}{100}$ b $\frac{51}{100}$, 0.67, 0.8 c $\frac{3}{10}$, $\frac{79}{100}$, 1.5
d $1\frac{4}{10}$, 1.56, $\frac{17}{10}$ e $\frac{79}{100}$, $\frac{4}{5}$, 0.9 f 0.08, $\frac{11}{100}$, 0.13
g $\frac{1}{2}$, 0.6, $\frac{33}{50}$ h 0.3, $\frac{31}{100}$, $\frac{7}{20}$ i $\frac{12}{20}$, 0.7, $\frac{3}{4}$
j $2\frac{37}{100}$, $2\frac{4}{5}$, 2.81 k $\frac{3}{10}$, $\frac{9}{25}$, 0.39 l 1.09, $1\frac{3}{5}$, 1.9
m $\frac{19}{100}$, 0.2, $\frac{1}{4}$ n $\frac{41}{100}$, $\frac{22}{50}$, 0.45 o 2.37, $\frac{12}{5}$, $2\frac{11}{25}$

Page 22

1 $\frac{31}{100}$, 0.31, 31% **2** $\frac{79}{100}$, 0.79, 79% **3** $\frac{17}{100}$, 0.17, 17%
4 $\frac{23}{100}$, 0.23, 23% **5** $\frac{5}{100}$ ($\frac{1}{20}$), 0.05, 5% **6** $\frac{95}{100}$ ($\frac{19}{20}$), 0.95, 95%
7 $\frac{40}{100}$ ($\frac{2}{5}$), 0.4, 40% **8** $\frac{64}{100}$ ($\frac{16}{25}$), 0.64, 64% **9** $\frac{19}{100}$, 0.19, 19%
10 $\frac{36}{100}$ ($\frac{9}{25}$), 0.36, 36% **11** $\frac{8}{100}$ ($\frac{2}{25}$), 0.08, 8% **12** $\frac{90}{100}$ ($\frac{9}{10}$), 0.9, 90%
13 $\frac{47}{100}$, 0.47, 47% **14** $\frac{50}{100}$ ($\frac{1}{2}$), 0.5, 50% **15** $\frac{32}{100}$ ($\frac{8}{25}$), 0.32, 32%
16 $\frac{12}{100}$ ($\frac{3}{25}$), 0.12, 12%

Pages 23–24

1 a $\frac{7}{10}$ b $\frac{81}{100}$ c $\frac{11}{20}$ d $\frac{3}{10}$ e $\frac{12}{25}$ f $\frac{3}{20}$
g $\frac{3}{100}$ h $\frac{13}{50}$ i $\frac{18}{25}$ j $\frac{16}{25}$ k $\frac{37}{100}$ l $\frac{2}{25}$
m $\frac{3}{25}$ n $\frac{9}{10}$ o $\frac{1}{2}$ p $\frac{41}{100}$ q $\frac{3}{4}$ r $\frac{67}{100}$
s $\frac{33}{100}$ t $\frac{59}{100}$ u $\frac{6}{25}$ v $\frac{99}{100}$ w $\frac{1}{5}$ x $\frac{1}{10}$

2 a $\frac{37}{100}$, $\frac{8}{100}$, $\frac{33}{100}$, $\frac{46}{100}$, $\frac{79}{100}$, $\frac{85}{100}$, $\frac{17}{100}$, $\frac{61}{100}$
b $\frac{15}{20}$, $\frac{7}{25}$, $\frac{4}{5}$, $\frac{9}{50}$, $\frac{9}{10}$, $\frac{2}{5}$, $\frac{11}{25}$, $\frac{1}{10}$, $\frac{8}{20}$, $\frac{3}{4}$, $\frac{4}{10}$, $\frac{28}{50}$, $\frac{20}{25}$

3 a 39% b 21% c 45% d 13% e 67% f 7%
g 10% h 90% i 20% j 68% k 32% l 80%
m 30% n 60% o 24% p 45% q 84% r 20%
s 44% t 85% u 75% v 50% w 25% x 100%

4 a 0.25 b 0.1 c 0.5 d 0.75 e 0.2 f 0.17
g 0.39 h 0.95 i 0.72 j 0.61 k 0.84 l 0.42
m 0.07 n 0.04 o 0.3 p 0.48 q 0.93 r 0.11
s 0.7 t 0.35 u 0.01 v 0.22 w 0.53 x 0.09

5 a 57% b 34% c 65% d 23% e 86% f 14%
g 92% h 2% i 74% j 5% k 41% l 59%
m 80% n 30% o 50% p 10% q 40% r 90%
s 77% t 18% u 95% v 38% w 25% x 75%

6

a $\frac{54}{100}$ = 0.54 = 54%, 0.39 = $\frac{39}{100}$ = 39%, 22% = $\frac{22}{100}$ = 0.22, $\frac{86}{100}$ = 0.86 = 86%, 0.7 = $\frac{7}{10}$ = 70%, 5% = $\frac{5}{100}$ = 0.05, $\frac{1}{2}$ = 0.5 = 50%, 0.31 = $\frac{31}{100}$ = 31%

b 0.08 = $\frac{8}{100}$ = 8%, $\frac{7}{20}$ = 0.35 = 35%, 0.5 = $\frac{5}{10}$ = 50%, 75% = $\frac{75}{100}$ = 0.75, $\frac{1}{4}$ = 0.25 = 25%, 40% = $\frac{40}{100}$ = 0.4, 0.55 = $\frac{55}{100}$ = 55%, $\frac{2}{5}$ = 0.4 = 40%

c $\frac{8}{25}$ = 0.32 = 32%, 10% = $\frac{10}{100}$ = 0.1, 0.03 = $\frac{3}{100}$ = 3%, 100% = $\frac{100}{100}$ = 1.0, $\frac{9}{10}$ = 0.9 = 90%, $\frac{3}{50}$ = 0.06 = 6%, 0.91 = $\frac{91}{100}$ = 91%, 0.2 = $\frac{2}{10}$ = 20%

d 67% = $\frac{67}{100}$ = 0.67, 0.8 = $\frac{8}{10}$ = 80%, $\frac{1}{5}$ = 0.2 = 20%, 2% = $\frac{2}{100}$ = 0.02, 16% = $\frac{16}{100}$ = 0.16, 0.15 = $\frac{15}{100}$ = 15%, $\frac{19}{20}$ = 0.95 = 95%, 0.43 = $\frac{43}{100}$ = 43%

e 0.72 = $\frac{72}{100}$ = 72%, $\frac{27}{50}$ = 0.54 = 54%, 30% = $\frac{3}{10}$ = 0.3, 0.44 = $\frac{44}{100}$ = 44%, $\frac{47}{100}$ = 0.47 = 47%, 96% = $\frac{96}{100}$ = 0.96, $\frac{16}{25}$ = 0.64 = 64%, 0.36 = $\frac{36}{100}$ = 36%

f $\frac{6}{10}$ = 0.6 = 60%, 0.13 = $\frac{13}{100}$ = 13%, 57% = $\frac{57}{100}$ = 0.57, 0.09 = $\frac{9}{100}$ = 9%, $\frac{14}{20}$ = 0.7 = 70%, 29% = 0.29 = $\frac{29}{100}$, 0.33 = $\frac{33}{100}$ = 33%, 87% = 0.87 = $\frac{87}{100}$

g 41% = $\frac{41}{100}$ = 0.41, $\frac{79}{100}$ = 0.79 = 79%, 0.99 = $\frac{99}{100}$ = 99%, $\frac{9}{20}$ = 0.45 = 45%, 23% = $\frac{23}{100}$ = 0.23, 0.51 = $\frac{51}{100}$ = 51%, 83% = 0.83 = $\frac{83}{100}$, $\frac{3}{4}$ = 0.75 = 75%

Page 25

1

	%	Common fraction	Diagram	Number line	Decimal
a	25%	$\frac{1}{4}$		0 ↓ 1	0.25
b	20%	$\frac{1}{5}$	Diagrams will vary but must show $\frac{1}{5}$	0 ↓ 1	0.2
c	50%	$\frac{1}{2}$		0 ↓ 1	0.5
d	75%	$\frac{3}{4}$	Diagrams will vary but must show $\frac{3}{4}$	0 ↓ 1	0.75
e	100%	$\frac{1}{1}$		0 ↓1	1.0
f	10%	$\frac{1}{10}$	Diagrams will vary but must show $\frac{1}{10}$	0 ↓ 1	0.1

2 a 3.5 b 0.235 c 0.57 d $\frac{71}{100}$ e $\frac{3}{10}$ f $\frac{37}{100}$
g 1.6 h 95% i $\frac{1}{2}$ j 35% k $\frac{3}{4}$ l $\frac{5}{10}$
m $\frac{57}{100}$ n 0.88 o 0.65 p 4.1 q 0.93 r 1.5
s 3.6 t $\frac{15}{50}$

3 a = b < c < d < e > f =
g = h > i = j < k > l >
m > n < o < p <

Page 26

1 a 10 b 45 c 9 d 56 e 3 f 36
g 55 h 5 i 250 j 03 k 96 l 8
m 3 n 18 o 26 p 27 q 7 r 10
s 12 t 3 u 60 v 6 w 36 x 162

2 a 5.7 b 9.5 c 5.25 d 3.6 e 12.75 f 8.6
g 14.4 h 42.5 i 2.1 j 15.75 k 21.75 l 6.4

3 a K5 b K10 c K15 d K1 e K1.50 f K9
g K30 h K2 i K7.50 j K32

4 a K20 b K95 c K54 d K375 e K105 f K40.50
g K26.60 h K52.50 i K28.80 j K102

Page 27

1 a 3 × 3 b 5 × 5 × 5 c 4 × 4 d 1 × 1 × 1
e 9 × 9 f 6 × 6 × 6 g 8 × 8 × 8 h 10 × 10 × 10
i 7 × 7 j 12 × 12 k 15 × 15 × 15 l 11 × 11
m 5 × 5 n 2 × 2 o 3 × 3 × 3 p 9 × 9 × 9
q 10 × 10 r 4 × 4 × 4

2 a 9^3 b 7^3 c 6^2 d 3^3 e 13^2 f 2^2
g 10^2 h 18^2 i 8^3 j 2^3 k 14^2 l 21^2
m 3^2 n 1^3 o 5^3 p 11^2 q 9^2 r 16^2

3 a 16 b 25 c 100 d 64 e 225 f 400
g 8 h 1 i 125 j 1000 k 64 l 343
m 81 n 216 o 144 p 729 q 900 r 8000

4 a 54 b 64 c 2000 d 200 e 250 f 448
g 9 h 324 i 108 j 320 k 72 l 6000
m 128 n 81 o 40 p 490 q 432 r 432

5 a 19 b 68 c 1008 d 76 e 52 f 24
g 18 h 225 i 35 j 0 k 80 l 135
m 54 n 900 o 188

6 a < b > c < d > e > f <
g < h = i > j > k = l <
m < n > o <

Page 28 Assessment

1 C **2** B **3** C **4** B **5** A
6 D **7** C **8** B **9** C **10** C
11 A **12** B **13** D **14** B **15** A
16 D **17** C **18** D **19** B **20** B

Pages 29–31 Assessment

1 **a** 42.63 **b** 93.75 **c** 25.255 **d** 192.12
e 22.71 **f** 23.79 **g** 23.775 **h** 48.26

2 **a** 45.5 **b** 74.7 **c** 34.2 **d** 38.4 **e** 71.4 **f** 155.2
g 184.5 **h** 316.4 **i** 27.32 **j** 29.76 **k** 103.95 **l** 291.92

3 **a** 3.2 **b** 2.3 **c** 11.2 **d** 3.23 **e** 1.3 **f** 1.4
g 4.4 **h** 5.5 **i** 0.459 **j** 0.716 **k** 8.87 **l** 5.88

4 **a** 0.7 **b** 0.59 **c** 0.34 **d** 0.1 **e** 0.8 **f** 0.68
g 0.18 **h** 0.46 **i** 0.35 **j** 0.55 **k** 0.3 **l** 0.4

5 **a** $\frac{9}{10}$ **b** $\frac{81}{100}$ **c** $\frac{4}{5}$ **d** $\frac{6}{25}$ **e** $\frac{14}{25}$ **f** $\frac{21}{50}$
g $2\frac{1}{2}$ **h** $4\frac{3}{4}$ **i** $1\frac{3}{50}$ **j** $9\frac{3}{5}$ **k** $11\frac{1}{4}$ **l** $4\frac{3}{20}$

6 **a** $\frac{4}{5}$ **b** $\frac{1}{4}$ **c** $\frac{1}{10}$ **d** $\frac{19}{50}$ **e** $\frac{51}{100}$ **f** $\frac{4}{25}$
g $\frac{63}{100}$ **h** $\frac{24}{25}$ **i** $\frac{1}{25}$ **j** $\frac{37}{50}$ **k** $\frac{11}{50}$ **l** $\frac{9}{20}$

7 **a** 31% **b** 6% **c** 18% **d** 54% **e** 35% **f** 36%
g 90% **h** 40% **i** 75% **j** 80% **k** 88% **l** 90%

8 **a** 0.98 **b** 0.03 **c** 0.13 **d** 0.58 **e** 0.41 **f** 0.76
g 0.11 **h** 0.29 **i** 0.62 **j** 0.35 **k** 0.84 **l** 0.9

9 **a** 61% **b** 40% **c** 7% **d** 55% **e** 89% **f** 36%
g 20% **h** 97% **i** 14% **j** 76% **k** 23% **l** 42%

10 **a** 49% **b** 72% **c** $\frac{2}{5}$ **d** $\frac{25}{100}$ **e** 80% **f** $\frac{5}{10}$
g 10% **h** 1.0 **i** 85% **j** 6.3 **k** $\frac{14}{25}$ **l** $\frac{11}{20}$

11 **a** $\frac{75}{100}$, 0.7, 7% **b** $\frac{5}{20}$, 20%. 0.05 **c** $\frac{4}{10}$, 37%, 0.3
d 46%, $\frac{4}{100}$, 0.046 **e** $\frac{9}{10}$, 87%, 0.85 **f** 0.7, $\frac{1}{5}$, 17%
g 90%, 0.09, $\frac{3}{50}$ **h** 36%, 0.35, $\frac{1}{3}$ **i** 0.65, 63%, $\frac{3}{5}$
j 0.55, $\frac{26}{50}$, 5% **k** $\frac{19}{20}$, 0.92, 90% **l** $\frac{1}{4}$, 20%, 0.02

12 **a** 16 **b** 36 **c** 27 **d** 17 **e** 17 **f** 24
g 2 **h** 15 **i** 56 **j** 90 **k** 3 **l** 54

13 **a** 10.5 **b** 2.25 **c** 8.7 **d** 7.6 **e** 24.75 **f** 1.3
g 27.5 **h** 11.25 **i** 7.2 **j** 12.6 **k** 36.75 **l** 2.9

14 **a** 20×20 **b** $4 \times 4 \times 4$ **c** 8×8 **d** $15 \times 15 \times 15$
e 25×25 **f** $12 \times 12 \times 12$ **g** $18 \times 18 \times 18$ **h** 50×50
i $24 \times 24 \times 24$ **j** 54×54 **k** $75 \times 75 \times 75$ **l** 63×63

15 **a** 14^2 **b** 30^2 **c** 19^3 **d** 11^3 **e** 26^2 **f** 45^3
g 17^2 **h** 32^3

16 **a** 48 **b** 48 **c** 250 **d** 2000 **e** 72 **f** 89
g 373 **h** 280 **i** 144 **j** 1000 **k** 80 **l** 1343

17 **a** 456 552 **b** 60 208 **c** 22 576 **d** 1373
e 89 681 **f** 342 846

Strand: Space and Shape

Pages 32–34

1 **a** 700 cm **b** 300 cm **c** 550 cm **d** 850 cm
e 325 cm **f** 475 cm **g** 310 cm **h** 190 cm
i 630 cm **j** 180 cm **k** 220 cm **l** 70 cm

2 **a** 50 mm **b** 35 mm **c** 100 mm **d** 25 mm
e 120 mm **f** 300 mm **g** 150 mm **h** 85 mm
i 250 mm **j** 360 mm **k** 185 mm **l** 1000 mm

3 **a** 3 cm **b** 7 cm **c** 1 cm **d** 15 cm
e 11 cm **f** 9 cm **g** 20 cm **h** 18 cm
i 50 cm **j** 35 cm **k** 48 cm **l** 100 cm

4 **a** 5 m **b** 12 m **c** 8 m **d** 20 m
e 36 m **f** 14 m **g** 59 m **h** 41 m
i 85 m **j** 100 m **k** 120 m **l** 116 m

5 **a** 7000 mm **b** 12 000 mm **c** 8000 mm **d** 2000 mm
e 4500 mm **f** 10 500 mm **g** 1250 mm **h** 5250 mm
i 3750 mm **j** 6750 mm **k** 900 mm **l** 11 700 mm

6 **a** 3 m **b** 5 m **c** 14 m **d** 11 m **e** $6\frac{1}{2}$ m **f** $1\frac{1}{2}$ m
g $12\frac{1}{2}$ m **h** $1\frac{1}{4}$ m **i** $7\frac{1}{4}$ m **j** $9\frac{3}{4}$ m **k** $\frac{3}{4}$ m **l** $\frac{1}{4}$ m

7 **a** 5000 m **b** 10 500 m **c** 4250 m **d** 7750 m
e 200 m **f** 900 m **g** 1300 m **h** 25 000 m
i 11 250 m **j** 8800 m **k** 6700 m **l** 12 125 m

8 **a** 4 km **b** 9 km **c** 15 km **d** 22 km
e $2\frac{1}{2}$ km **f** $11\frac{1}{2}$ km **g** $40\frac{1}{2}$ km **h** $\frac{1}{4}$ km
i $5\frac{1}{4}$ km **j** $7\frac{3}{4}$ km **k** $16\frac{3}{4}$ km **l** $13\frac{1}{5}$ km

9 **a** 206 cm **b** 140 cm **c** 545 cm **d** 70 cm
e 318 cm **f** 480 cm **g** 107 cm **h** 665 cm
i 138 cm **j** 1502 cm **k** 1150 cm **l** 1048 cm

10 **a** 2.43 m **b** 1.67 m **c** 0.79 m **d** 3.16 m
e 5.4 m **f** 1.12 m **g** 0.33 m **h** 2.89 m
i 8.04 m **j** 9.5 m **k** 3.11 m **l** 7.65 m

11 **a** 24 mm **b** 47 mm **c** 13 mm **d** 85 mm
e 38 mm **f** 112 mm **g** 56 mm **h** 129 mm
i 8 mm **j** 105 mm **k** 1 mm **l** 158 mm

12 **a** 12.9 cm **b** 41.3 cm **c** 30.5 cm **d** 7.8 cm
e 56.5 cm **f** 5.3 cm **g** 23.7 cm **h** 4.1 cm
i 47.2 cm **j** 13.4 cm **k** 26 cm **l** 8 cm

13 **a** 1800 m **b** 4600 m **c** 7900 m **d** 15 100 m
e 300 m **f** 3250 m **g** 8370 m **h** 5020 m
i 11 790 m **j** 1672 m **k** 6015 m **l** 916 m

14 **a** 2.8 km **b** 5.9 km **c** 8.7 km **d** 3.45 km
e 1.375 km **f** 1.059 km **g** 0.925 km **h** 0.709 km
i 10.625 km **j** 12.004 km **k** 18.3 km **l** 11.61 km

15 **a** 1500 mm **b** 3900 mm **c** 400 mm **d** 9700 mm
e 12 600 mm **f** 10 300 mm **g** 6250 mm **h** 4080 mm
i 7750 mm **j** 11 020 mm **k** 4176 mm **l** 978 mm

16 **a** 3.1 m **b** 2.8 m **c** 6.3 m **d** 9.85 m
e 5.076 m **f** 8.109 m **g** 0.635 m **h** 0.884 m
i 17.865 m **j** 23.2 m **k** 31.09 m **l** 12.008 m

Pages 34–35

1 a 785 cm b 51785 cm c 2825 m d 847 cm
e 4675 cm f 6080 m g 1410 cm h 89 mm
i 27500 m j 1695 cm k 28840 cm l 109 mm
m 1008 cm n 2055 mm o 1960 cm p 4761 m
q 641 mm r 844 cm s 7930 m t 5480 m

2 a 1 m b 6 m c 4 m d 1 m
e 2 m f 3 m g 15 m h 17 m
i 31 m j 10 m k 2 m l 8 m
m 9 m n 2 m o 7 m p 4 m
q 8 m r 4 m s 27 m t 3 m

3 a 4 cm b 2 cm c 52 cm d 33 cm e 5 cm f 86 cm
g 5 cm h 19 cm i 78 cm j 102 cm k 13 cm l 8 cm
m 12 cm n 2 cm o 58 cm p 21 cm q 11 cm r 65 cm
s 717 cm t 503 cm

4 a 9 km b 16 km c 1 km d 5 km e 1 km f 7 km
g 7 km h 11 km i 28 km j 18 km k 32 km l 24 km
m 1 km n 4 km o 9 km p 6 km q 12 km r 2 km
s 1 km t 20 km

5 a, b, d, f, g and h can be rounded off to 7 metres

6 a, c, f, h and j can be rounded off to 10 kilometres

7 b, d, e, g, h and j can be rounded off to 6 centimetres

8

	Length	Nearest metre	Nearest kilometre
a	725.4 m	725 m	1 km
b	908.31 m	908 m	1 km
c	655.9 m	656 m	1 km
d	1843.7 m	1844 m	2 km
e	3065.2 m	3065 m	3 km
f	2689.6 m	2690 m	3 km
g	1516.78 m	1517 m	2 km
h	5918.49 m	5918 m	6 km
i	8235.8 m	8236 m	8 km
j	4704.63 m	4705 m	5 km

9

	Length	Nearest centimetre	Nearest metre
a	587 mm	59 cm	1 m
b	868 mm	87 cm	1 m
c	635 mm	64 cm	1 m
d	1588 mm	159 cm	2 m
e	1925 mm	193 cm	2 m
f	3154 mm	315 cm	3 m
g	5936 mm	594 cm	6 m
h	3781 mm	378 cm	4 m
i	8158 mm	816 cm	8 m
j	4712 mm	471 cm	5 m

Page 36

1 a 8 km b 10 km c 9 km d 11 km e 12 km f 12 km

2 a 13 cm b 10 cm c 13 cm d 13.5 cm e 12 cm f 15 cm

Page 37

1 a 11 m b 10 m c 9 m d 14.5 km e 14.5 km f 12.5 m
g 21 km h 34 m

2 a 10 cm b 13 cm c 13 cm d 10 cm e 13 cm f 13 cm
g 16 cm h 23 cm

Page 38

1 a 32 cm^2 b 4 m^2 c 48 cm^2 d 10 cm^2 e 96 cm^2 f 25 m^2
g 48 cm^2 h 12 cm^2 i 40 m^2 j 15 cm^2 k 21 m^2

2 a 35 cm^2 b 6 cm c 5 cm d 6 cm e 40 cm^2 f 4 cm
g 54 cm^2 h 5 cm i 12 cm j 9 cm k 72 cm^2 l 5 cm

3 a 1 cm × 36 cm, 2 cm × 18 cm, 3 cm × 12 cm, 4 cm × 9 cm, 6 cm × 6 cm
b 1 cm × 16 cm, 2 cm × 8 cm, 4 cm × 4 cm
c 1 cm × 40 cm, 2 cm × 20 cm, 4 cm × 10 cm, 5 cm × 8 cm
d 1 cm × 18 cm, 2 cm × 9 cm, 3 cm × 6 cm

Page 39

1 15 cm^2 **2** 8 cm^2 **3** 17.5 cm^2 **4** 21 cm^2 **5** 25 cm^2
6 22 cm^2 **7** 16 cm^2 **8** 20 cm^2 **9** 60 cm^2 **10** 13 cm^2
11 32.5 cm^2 **12** 10.5 cm^2 **13** 90 cm^2 **14** 17 cm^2 **15** 37.5 cm^2

Page 40

1 32 m^2 **2** 27 m^2 **3** 85 m^2 **4** 28 m^2 **5** 49 m^2
6 46 m^2 **7** 28 m^2 **8** 23 m^2 **9** 47 m^2 **10** 40 m^2
11 12 m^2 **12** 52 m^2 **13** 44 m^2 **14** 55 m^2

Page 41

1 20 cm^3 **2** 60 cm^3 **3** 36 cm^3 **4** 32 cm^3 **5** 32 cm^3
6 72 cm^3 **7** 80 cm^3 **8** 12 cm^3 **9** 24 cm^3 **10** 84 cm^3
11 45 cm^3 **12** 9 cm^3 **13** 60 cm^3 **14** 45 cm^3

Page 42

1 100 cm^3 **2** 50 cm^3 **3** 105 cm^3 **4** 264 cm^3 **5** 60 cm^3
6 150 cm^3 **7** 56 cm^3 **8** 180 cm^3 **9** 54 cm^3 **10** 27 cm^3
11 30 cm^3 **12** 75 cm^3

Page 43

1 a 1560 mL b 3450 mL c 2125 mL d 5615 mL
e 2975 mL f 10723 mL g 4364 mL h 8045 mL
i 12075 mL j 1030 mL k 3005 mL l 6008 mL

2 a 5 L 641 mL b 2 L 200 mL c 1 L 759 mL d 3 L 85 mL
e 1 L 60 mL f 7 L 5 mL g 0 L 817 mL h 0 L 384 mL
i 12 L 356 mL j 10 L 95 mL k 15 L 208 mL l 11 L 410 mL

3 a 1500 mL b 2750 mL c 200 mL d 1250 mL
e 3500 mL f 5625 mL g 815 mL h 3270 mL
i 1600 mL j 7360 mL k 9700 mL l 6050 mL
m 3619 mL n 2800 mL o 5500 mL p 429 mL
q 1987 mL r 250 mL s 2311 mL t 80 mL
u 1340 mL v 7060 mL w 3088 mL x 4390 mL

4 a 2.5 L b 4.5 L c 1.75 L d 0.465 L
e 0.275 L f 1.075 L g 0.218 L h 0.032 L
i 1.8 L j 1.03 L k 0.41 L l 5.4 L
m 0.875 L n 2.3 L o 4.01 L p 0.075 L
q 0.04 L r 5.109 L s 0.633 L t 3.002 L

5 a 5500 mL b 250 mL c 1250 mL d 500 mL
e 750 mL f 3750 mL g 500 mL h 900 mL
i 4200 mL j 2250 mL k 1300 mL l 2750 mL
m 7250 mL n 200 mL o 4800 mL p 125 mL
q 3125 mL r 200 mL

6 a $2\frac{1}{2}$ L b $3\frac{1}{4}$ L c $\frac{3}{4}$ L d $4\frac{1}{5}$ L e $\frac{9}{10}$ L f $5\frac{1}{10}$ L
g $6\frac{1}{2}$ L h $1\frac{3}{4}$ L i $3\frac{1}{2}$ L j $\frac{1}{4}$ L k $\frac{1}{8}$ L l $5\frac{1}{8}$ L
m $7\frac{3}{4}$ L n $2\frac{3}{5}$ L o $\frac{1}{2}$ L p $\frac{1}{10}$ L q $1\frac{1}{4}$ L r $6\frac{1}{8}$ L
s $8\frac{1}{10}$ L t $\frac{5}{8}$ L

Page 44

1 a 60 000 cm^2 b 20 000 cm^2 c 70 000 cm^2 d 30 000 cm^2
e 15 000 cm^2 f 95 000 cm^2 g 55 000 cm^2 h 5000 cm^2
i 100 000 cm^2 j 82 500 cm^2 k 42 500 cm^2 l 12 500 cm^2
m 57 500 cm^2 n 27 500 cm^2 o 7500 cm^2 p 107 500 cm^2
q 38 000 cm^2 r 72 000 cm^2 s 19 000 cm^2 t 61 250 cm^2
u 3750 cm^2 v 26 250 cm^2 w 10 750 cm^2 x 90 250 cm^2

2 a 2 m^2 b 5 m^2 c 17 m^2 d 24 m^2
e 2.5 m^2 f 7.5 m^2 g 4.5 m^2 h 1.5 m^2
i 3.05 m^2 j 1.85 m^2 k 1.523 m^2 l 2.718 m^2
m 0.85 m^2 n 0.589 m^2 o 4.3054 m^2 p 7.514 m^2
q 10.265 m^2 r 34.008 m^2 s 0.3005 m^2 t 0.967 m^2

3 a 40 000 m^2 b 70 000 m^2 c 150 000 m^2 d 110 000 m^2
e 85 000 m^2 f 35 000 m^2 g 105 000 m^2 h 15 000 m^2
i 32 500 m^2 j 122 500 m^2 k 57 500 m^2 l 97 500 m^2
m 62 000 m^2 n 48 000 m^2 o 117 000 m^2 p 29 000 m^2
q 3000 m^2 r 81 250 m^2 s 76 250 m^2 t 670 m^2
u 29 700 m^2 v 163 500 m^2 w 10 080 m2 x 35 040 m^2

4 a 2 ha b 8 ha c 13 ha d 1.5 ha
e 6.5 ha f 23.5 ha g 6.25 ha h 3.75 ha
i 8.35 ha j 4.925 ha k 5.175 ha l 7.075 ha
m 4.865 ha n 2.705 ha o 12.628 ha p 10.07 ha
q 21.06 ha r 8.89 ha s 0.9 ha t 0.725 ha

5 a 32 500 m^2 b 27 500 cm^2 c 76 000 m^2 d 52 500 cm^2
e 5000 m^2 f 65 000 m^2 g 7000 m^2 h 44 000 cm^2
i 102 500 m^2 j 8000 cm^2 k 19 000 m^2 l 21 250 cm^2
m 53 750 m^2 n 71 250 cm^2 o 46 250 m^2 p 8750 cm^2
q 30 500 m^2 r 80 200 cm^2 s 116 500 m^2 t 66 800 cm^2
u 54 500 m^2

Page 45

1 a acute b straight c right d obtuse
e reflex f obtuse g acute h reflex
i obtuse j right k straight l obtuse
m acute n acute o reflex p obtuse

2 a acute b reflex c obtuse d obtuse
e obtuse f right g right h acute
i reflex j obtuse k right l obtuse

Page 46

1 90° **2** 40° **3** 20° **4** 140° **5** 60°
6 100° **7** 30° **8** 80° **9** 120° **10** 90°
11 110° **12** 25° **13** 70° **14** 50° **15** 125°
16 10° **17** 150° **18** 55° **19** 45° **20** 130°

Page 47

1 120° **2** 30° **3** 110° **4** 90° **5** 45°
6 70° **7** 30° **8** 35° **9** 270° **10** 330°
11 240° **12** 280° **13** 35° **14** 60° **15** 80°
16 55° **17** 25° **18** 50° **19** 100° **20** 120°

Page 48

1 a 3 cm b 5 cm c 10 cm d 1.5 cm e 0.5 cm f 2.5 cm
g 6.5 cm h 25 cm i 5.5 cm j 13.5 cm k 7.5 cm l 19.5 cm

2 a 2 cm b 4 cm c 10 cm d 7 cm e 0.5 cm f 3 cm
g 4.5 cm h 7.5 cm i 2.5 cm j 13.5 cm k 16.5 cm l 9 cm

3 a 2 cm b 4 cm c 0.5 cm d 10 cm e 5 cm f 2.5 cm
g 1.5 cm h 4.5 cm i 10.5 cm j 8.5 cm k 3.5 cm l 12.5 cm

4 a 3 cm b 0.5 cm c 10 cm d 20 cm e 1.5 cm f 2.5 cm
g 7 cm h 10.5 cm i 4.5 cm j 7.5 cm k 8.5 cm l 21.5 cm

5 a 200 km b 50 km c 25 km d 500 km e 350 km f 750 km
g 400 km h 125 km i 325 km j 175 km k 475 km l 575 km

6 a 100 km b 200 km c 150 km d 500 km e 25 km f 20 km
g 10 km h 300 km i 250 km j 400 km k 350 km l 15 km

7 a 200 km b 40 km c 60 km d 160 km e 100 km f 10 km
g 30 km h 90 km i 400 km j 120 km k 150 km l 70 km

Page 49

Students are required to play a game.

Pages 50–53 Assessment

1 a 450 cm b 540 cm c 75 mm d 260 mm
e 42 cm f 8.5 cm g 9 m h 53 m
i 6000 mm j 4750 mm k $8\frac{1}{2}$ m l $3\frac{1}{4}$ m
m 9000 m n 2700 m o $5\frac{1}{2}$ km p $\frac{3}{4}$ km
q 920 cm r 602 cm s 78 mm t 116 mm
u 8500 m v 600 m w 2700 mm x 5030 mm

2 a 800 cm b 177 530 cm or 1775.3 m c 116 mm d 628 cm

3 a 8 m b 27 m c 11 m d 15 m e 16 m

4 a 6 cm b 32 cm c 404 cm d 25 cm e 16 cm

5 a 11 km b 32 km c 1 km d 3 km e 29 km

6 a 16 m b 17 m c 17 m d 32 m

7 a 10 cm^2 b 63 cm^2 c 36 cm^2 d 24 cm^2

8 a 24 cm^2 b 10 cm^2 c 55 cm^2 d 18 cm^2

9 a 32 m^2 b 72 m^2 c 46.5 m^2 d 92 m^2

10 a 24 cm^3 b 120 cm^3 c 20 cm^3 d 60 cm^3

11 a 140 m^3 b 24 m^3 c 20 m^3 d 72 m^3

12 a 8265 mL b 7750 mL c 5200 mL d 2015 mL e 860 mL f 3875 mL g 6050 mL h 4400 mL i 10 008 mL j 9340 mL k 5350 mL l 1900 mL m 3422 mL n 7125 mL o 18 000 mL

13 a 100 000 cm^2 b 50 000 m^2 c 27 500 cm^2 d 73 000 m^2 e 68 000 cm^2 f 40 020 m^2 g 31 250 cm^2 h 118 750 m^2 i 3000 cm^2 j 13 550 m^2 k 78 000 cm^2 l 500 m^2

14 a acute b obtuse c reflex d straight e acute

15 a 110° b 90° c 60° d 100°

16 a 130° b 320° c 45° d 70° e 70°

17 a 3 cm b 0.5 cm c 8 cm d 10 cm e 5 cm f 2.5 cm g 1.5 cm h 4.5 cm i 20 cm j 7.5 cm k 3.5 cm l 9 cm m 2 cm n 12.5 cm o 9.5 cm p 5.5 cm q 7 cm r 15 cm

18 a 240 km b 1200 km c 60 km d 40 km e 280 km f 600 km g 80 km h 760 km i 960 km j 540 km k 440 km l 160 km m 480 km n 1040 km o 2400 km p 800 km q 360 km r 1320 km

Strand: Measurement

Pages 54–55

1 315 g, 0.25 kg, 600 g, 570 g, 275 g, 0.68 kg, 375 g, 0.5 kg

2 2.06 kg, 1.8 kg, 2.2 kg, 2375 g, 1990 g, 2.04 kg, 1.95 kg, 2250 g, 1.76 kg

3 950 g, 0.9 kg, 805 g, 1 kg, 1.09 kg, 1.245 kg, 1.035 kg, 860 g, 0.85 kg, 1.12 kg, 1.23 kg

4 310 g, 315 g, 0.316 kg, 0.33 kg

5 a 0.7 kg b 1.5 kg c 405 g d 3.1 kg e 2.8 kg f 815 g g 0.55 kg h 0.275 kg i 675 g j 3800 g k 755 g l 0.6 kg m 1.1 kg n $\frac{1}{4}$ kg o 4150 g p 710 g q 0.8 kg r 0.95 kg

6 a–f Teacher to check

Pages 55–56

1 a 1500 g b 2750 g c 695 g d 1200 g e 4350 g f 636 g g 178 g h 3080 g i 2195 g j 1056 g k 5950 g l 372 g m 5070 g n 5 g o 2106 g p 1090 g q 9120 g r 5970 g s 3100 g t 87 g u 2050 g v 441 g w 1703 g x 3790 g

2 a 2 kg b 1.5 kg c 7 kg d 5.5 kg e 3.25 kg f 1.75 kg g 4.25 kg h 8.75 kg i 6.7 kg j 1.85 kg k 1.52 kg l 2.78 kg m 0.85 kg n 0.59 kg o 3.054 kg p 7.258 kg q 1.067 kg r 0.609 kg s 0.054 kg t 5.005 kg u 2.32 kg v 13.081 kg w 15.69 kg x 10.7 kg

3 a 7000 kg b 11 000 kg c 4500 kg d 6250 kg e 3250 kg f 7050 kg g 15 100 kg h 3067 kg i 2400 kg j 134 kg k 5750 kg l 8700 kg m 120 kg n 1088 kg o 4750 kg p 3875 kg q 6970 kg r 4 kg s 2096 kg t 12 400 kg u 10 070 kg v 7058 kg w 99 kg x 1530 kg

4 a 9 t b 2 t c 13 t d 21 t e 6.5 t f 4.5 t g 9.25 t h 3.75 t i 5.268 t j 1.752 t k 2.085 t l 0.674 t m 8.6 t n 9.107 t o 0.51 t p 3.583 t q 4.59 t r 10.07 t s 1.316 t t 0.83 t u 12.158 t v 1.316 t w 0.38 t x 0.075 t

5 a 2250 g b 5750 kg c 3500 g d 7250 kg e 500 g f 1500 kg g 200 g h 3400 kg i 8300 g j 600 kg k 5700 g l 4125 kg m 2625 g n 6375 kg o 4875 g p 875 kg q 7125 g r 1050 kg s 9550 g t 4360 kg

Pages 56–57

1 a 250 kg b 750 kg c 1500 kg d 500 kg e 800 kg f 2400 kg g 3500 kg h 200 kg i 1875 kg j 1100 kg k 160 kg l 2040 kg

2 a 200 g b 750 g c 500 g d 400 g e 200 g f 3750 g g 300 g h 1250 g i 1400 g j 1200 g k 1600 g l 2250 g

3 a 2.4 t b 3 t c 2.94 t d 2.7 t e 5 t f 3.7 t g 13 t h 4.86 t i 5.39 t j 9.25 t k 3.48 t l 6.345 t

4 a 3.2 kg b 3.3 kg c 1.875 kg d 6.525 kg e 1.645 kg f 3.58 kg g 7.095 kg h 4.6 kg i 4.95 kg j 18.3 kg k 4.275 kg l 5.46 kg

5 a 8850 g b 4030 g c 2650 kg d 6405 g e 6317 kg f 2305 kg g 10 325 g h 1285 kg i 4705 g j 2150 g k 463 kg l 6633 g

6 a 355 g b 1.05 t c 7.8 kg d 6.525 kg e 11.4 kg f 118.75 kg g 36.89 kg h 973 g

Page 58

1 24°C **2** 37°C **3** 15°C **4** – 2°C **5** – 5°C **6** 36°C **7** 180°C **8** 0°C **9** 25°C **10** 49°C **11** – 5°C **12** 3°C

Page 59

1 $\frac{1}{2}$ past 6, 6:30 **2** 4 to 3, 2:56 **3** $\frac{1}{4}$ past 8, 8:15
4 20 to 4, 3:40 **5** 27 past 5, 5:27 **6** $\frac{1}{4}$ to 2, 1:45
7 $\frac{1}{2}$ past 11, 11:30 **8** 1 to 10, 9:59 **9** 8 past 4, 4:08
10 16 to 8, 7:44 **11** 3 to 11, 10:57 **12** 12 past 3, 3:12
13 9 to 6, 5:51 **14** 2 past 8, 8:02 **15** 29 to 3, 2:31

Pages 60–61

1 **a** 120 min **b** 300 min **c** 180 min **d** 600 min
e 90 min **f** 390 min **g** 135 min **h** 495 min
i 225 min **j** 105 min **k** 585 min **l** 345 min
m 140 min **n** 380 min **o** 40 min **p** 280 min
q 12 min **r** 432 min **s** 336 min **t** 108 min
u 486 min **v** 54 min **w** 162 min **x** 258 min

2 **a** 325 min **b** 190 min **c** 525 min **d** 170 min
e 95 min **f** 262 min **g** 558 min **h** 226 min
i 439 min **j** 117 min **k** 127 min **l** 332 min
m 204 min **n** 641 min **o** 413 min **p** 289 min
q 491 min **r** 327 min **s** 434 min **t** 86 min
u 159 min **v** 243 min **w** 703 min **x** 381 min

3 **a** 720 s **b** 240 s **c** 210 s **d** 450 s **e** 135 s **f** 345 s
g 620 s **h** 520 s **i** 264 s **j** 468 s **k** 96 s **l** 372 s
m 186 s **n** 534 s **o** 318 s **p** 162 s **q** 735 s **r** 705 s
s 250 s **t** 410 s

4 **a** 4 h 5 min **b** 2 h 10 min **c** 1 h 25 min **d** 5 h 20 min
e 8 h 20 min **f** 1 h 48 min **g** 2 h 26 min **h** 3 h 17 min
i 3 h 25 min **j** 6 h 53 min **k** 5 h 11 min **l** 4 h 29 min
m 10 h 24 min **n** 8 h 37 min **o** 1 h 13 min **p** 6 h 38 min
q 7 h 36 min **r** 9 h 12 min **s** 2 h 54 min **t** 3 h 46 min
u 5 h 39 min **v** 7 h 11 min **w** 8 h 28 min **x** 9 h 47 min

5 **a** 3 min 45 s **b** 1 min 34 s **c** 5 min 12 s **d** 9 min 26 s
e 1 min 53 s **f** 4 min 46 s **g** 7 min 41 s **h** 6 min 29 s
i 1 min 16 s **j** 2 min 38 s **k** 4 min 11 s **l** 3 min 15 s
m 5 min 43 s **n** 8 min 22 s **o** 6 min 50 s **p** 2 min 8 s
q 3 min 39 s **r** 6 min 36 s **s** 1 min 44 s **t** 10 min 27 s
u 13 min 46 s **v** 1 min 21 s **w** 4 min 23 s **x** 8 min 2 s

6 **a** 120 h **b** 48 h **c** 168 h **d** 96 h **e** 74 h **f** 59 h
g 137 h **h** 149 h **i** 244 h **j** 205 h **k** 81 h **l** 176 h
m 54 h **n** 156 h **o** 128 h **p** 90 h **q** 112 h **r** 171 h
s 87 h **t** 276 h

7 **a** 42 days **b** 30 days **c** 25 days **d** 61 days
e 43 days **f** 38 days **g** 72 days **h** 20 days
i 54 days **j** 34 days **k** 23 days **l** 84 days
m 105 days **n** 154 days **o** 133 days **p** 80 days
q 90 days **r** 122 days

8 **a** 24 months **b** 30 years **c** 60 months **d** 300 years
e 96 months **f** 60 years **g** 54 months **h** 700 years
i 27 months **j** 250 years **k** 55 years **l** 45 months
m 15 years **n** 425 years **o** 22 years **p** 80 months
q 96 years **r** 52 months **s** 20 decades **t** 35 decades
u 15 decades

9 **a** 2 days **b** $\frac{1}{2}$ day **c** 4 days **d** 10 days
e $\frac{1}{4}$ day **f** $1\frac{1}{2}$ days **g** 3 days **h** $3\frac{1}{2}$ days
i $1\frac{1}{4}$ days **j** 5 days **k** $\frac{3}{4}$ day **l** $2\frac{1}{2}$ days
m $1\frac{1}{8}$ day **n** $2\frac{1}{4}$ days **o** $\frac{2}{3}$ day **p** $2\frac{1}{3}$ days
q $4\frac{1}{3}$ days **r** $6\frac{2}{3}$ days

10 **a** 2 years **b** 5 years **c** $\frac{1}{2}$ year **d** 6 years
e $1\frac{1}{2}$ years **f** 3 years **g** $6\frac{1}{2}$ years **h** $1\frac{1}{3}$ years
i $2\frac{2}{3}$ years **j** $4\frac{1}{4}$ years **k** $8\frac{3}{4}$ years **l** $5\frac{1}{4}$ years
m $4\frac{2}{3}$ years **n** $9\frac{1}{3}$ years **o** 12 years **p** $7\frac{1}{2}$ years
q $3\frac{1}{3}$ years **r** $2\frac{3}{4}$ years **s** $10\frac{1}{4}$ years **t** $1\frac{1}{4}$ years

Page 62

1 **a** 1:20 p.m. **b** 8:43 p.m. **c** 4:19 p.m. **d** 8:27 a.m.
e 7:54 p.m. **f** 2:11 p.m. **g** 5:52 a.m. **h** 5:38 p.m.
i 10:05 p.m. **j** 9:42 a.m. **k** 10:56 a.m. **l** 11:16 p.m.
m 2:32 p.m. **n** 7:34 a.m. **o** 3:10 p.m. **p** 9 a.m.
q 5:06 p.m. **r** 1:45 p.m. **s** 2:41 a.m. **t** 6:59 p.m.
u 8:27 p.m. **v** 3:12 a.m. **w** 9:39 p.m. **x** 7:17 p.m.

2 **a** 1440 **b** 1724 **c** 1147 **d** 2206 **e** 1900 **f** 0639
g 2058 **h** 1516 **i** 0442 **j** 1821 **k** 1413 **l** 1208
m 1000 **n** 1215 **o** 0745 **p** 1355 **q** 2136 **r** 0550
s 2252 **t** 0605 **u** 2359 **v** 0045 **w** 1600 **x** 1530

3 **a** 2205, 2023, 1815 **b** 1550, 0859, 1412, 1148
c 0859 **d** 2205 **e** 1550 and 1735

4

	Morning/ afternoon	Clock	12-hour time	24-hour time
a	Morning		7:15 a.m.	0715
b	Afternoon	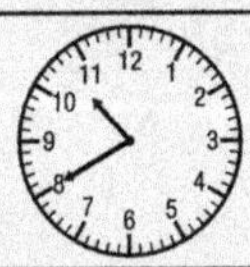	10:40 p.m.	2240
c	Afternoon		5:22 p.m.	1722
d	Afternoon		3:55 p.m.	1555
e	Afternoon		1.34 p.m.	1334

Page 63

1 a 19 min b 24 min c 52 min d 42 min e 38 min f 26 min
g 34 min h 24 min i 21 min j 28 min k 31 min l 25 min
m 51 min n 35 min o 35 min p 22 min q 36 min r 42 min
s 49 min t 27 min

2 a 7:17 p.m. b 4:59 p.m. c 10:39 a.m. d 12:14 p.m.
e 3:32 p.m. f 2:03 p.m. g 10:52 a.m. h 4:25 p.m.
i 3:06 p.m. j 12:29 p.m. k 10:10 a.m. l 8:35 p.m.
m 5:58 p.m. n 9:26 a.m. o 7:40 a.m. p 6:04 a.m.
q 11:28 a.m. r 5:46 p.m. s 8:54 p.m. t 10 p.m.
u 11:53 a.m. v 2:32 p.m. w 1:20 p.m. x 7:11 a.m.

3 a 1:52 p.m. b 9:32 a.m. c 11:02 a.m. d 2:32 p.m.
e 12:37 p.m. f 1:28 p.m. g 9:54 a.m. h 11:48 a.m.
i 3:30 p.m. j 12:46 p.m. k 4:34 p.m. l 2:40 p.m.
m 9:07 a.m. n 10:17 a.m. o 12 p.m. p 4:12 p.m.
q 4:27 p.m. r 2:09 p.m. s 10:34 a.m. t 12:51 p.m.
u 8:12 a.m. v 12:13 p.m. w 3:44 p.m. x 8:29 a.m.

4 a 5 h 35 min b 5 h 35 min c 6 h 20 min d 6 h 30 min
e 5 h 25 min f 3 h 50 min g 5 h 30 min h 4 h 35 min
i 6 h 5 min j 3 h 35 min k 6 h 55 min l 6 h 25 min
m 5 h 45 min n 7 h 20 min o 6 h 10 min p 6 h 25 min
q 5 h 40 min r 7 h 5 min

Page 64

1 Rani, John, Ruby, Karu, Lucy and Sam

2 Elli, Joseph, Mary, Julius, Dibara, Tau, Renaki, Charles and Mirou

3 Steven, Kila, Manu, Annette, Stanley, Elsie, Bernadette, Nalda, Ryan, Edward and Kate

4 Alena, Pulu, Kari, Sophie, Nora, Pia, Tapo, Amos, Jerry and Galema

Pages 65–68 Assessment

1 a 8 kg, 0.8 kg, 0.08 kg b 280 g, 0.25 kg, 0.2 kg
c $5\frac{3}{4}$ t, 5.6 t, 5500 kg d 4780 kg, 4.75 t, 4.075 t
e 2900 g, $2\frac{3}{4}$ kg, 2.7 kg f $3\frac{3}{8}$ t, 3125 kg, 3.01 t
g 0.4 t, 50 kg, 0.04 t h 2660 kg, $2\frac{3}{5}$ t, 2.06 t
i $7\frac{3}{10}$ kg, 7.1 kg, 710 g j 1.5 kg, $1\frac{2}{5}$ kg, 1050 g

2 a 800 g b 7500 g c 3950 g d 2375 g e 9030 g f 5400 g
g 10 750 g h 6 g i 2700 g j 11 650 g k 6125 g l 1170 g

3 a 1250 kg b 8200 kg c 875 kg d 105 kg e 4170 kg f 3680 kg
g 2920 kg h 5900 kg i 6030 kg j 7625 kg k 3900 kg l 1600 kg

4 a 5.6 t b 8.15 kg c 0.489 t d 9.002 kg e 11.025 t f 0.076 kg
g 3.208 t h 6.43 kg i 12.13 t j 5.81 kg k 0.092 t l 1.05 kg

5 a 500 kg b 1250 kg c 400 g d 2700 kg e 3750 g f 1500 kg
g 3600 g h 1600 g i 800 kg j 500 g k 250 g l 150 kg
m 100 g n 1125 kg o 1600 kg p 2800 g q 4375 kg r 1200 g
s 2800 kg t 2250 g

6 a 2.16 t b 1.72 kg c 2.7 t d 3.045 kg e 1.94 t f 2.064 kg
g 1.092 t h 0.912 kg i 0.7 kg j 0.525 t k 0.612 kg l 0.382 t
m 0.473 kg n 1.17 t o 2.86 kg p 0.13 t

7 a 1.34 t b 2.325 kg c 5.96 t d 4.446 kg e 0.751 kg f 2.143 kg
g 7.12 t h 7.175 t i 0.725 kg j 2.625 kg

8 a $\frac{1}{4}$ past 4, 4:15 b 26 to 11, 10:34 c 7 to 3, 2:53

9 a 195 min b 330 min c 140 min d 80 min e 45 min f 310 min
g 270 min h 222 min i 115 min j 144 min k 72 min l 275 min

10 a 570 s b 150 s c 320 s d 288 s e 405 s f 230 s
g 80 s h 228 s i 155 s j 460 s k 246 s l 365 s

11 a 1 h 26 min b 3 h 15 min c 1 h 50 min d 3 h 36 min
e 1 h 11 min f 5 h 38 min g 8 h 41 min h 4 h 52 min
i 6 h 47 min j 6 h 24 min

12 a 72 hours b 63 days c 48 months d 25 years
e 69 hours f 600 years g 59 days h 45 months
i 32 decades j 78 years k 92 months l 4 days
m 7 years n 30 decades o 12 minutes p 296 hours
q 570 years r 15 years

13 a 3:10 p.m. b 8:30 a.m. c 6:28 p.m. d 9:45 p.m.
e 11:50 p.m. f 3:40 a.m. g 10:17 a.m. h 1:52 p.m.
i 5:05 p.m. j 9:38 a.m. k 8:19 p.m. l 7:36 p.m.

14 a 0323 b 1441 c 1120 d 0756 e 1837 f 2009
g 0544 h 1618 i 1337 j 2148 k 1055 l 1204
m 0632 n 1453 o 0546

15 a 11:50 or 10 to 12 b 12:35 or 25 to 1 c 3:20 or 20 past 3
d 5:05 or 5 past 5 e 11:10 or 10 past 11 f 1:40 or 20 to 2
g 2:55 or 5 to 3 h 4:25 or 25 past 4

16 a 9:10 or 10 past 9 b 11:35 or 25 to 12 c 1:55 or 5 to 2
d 4:25 or 25 past 4

17 a 2:56 p.m. b 9:51 a.m. c 12:11 p.m. d 1:32 p.m.
e 5:13 p.m. f 11:02 a.m. g 5:43 p.m. h 1:18 p.m.
i 8:54 a.m. j 4:22 p.m.

18 a 1 h 40 min b 50 min c 1 h 40 min d 2 h 25 min
e 3 h 35 min f 3 h 15 min g 4 h 25 min h 1 h 35 min
i 2 h 35 min j 2 h 45 min k 1 h 47 min l 1 h 54 min
m 1 h 31 min n 2 h 38 min o 2 h 2 min p 2 h 21 min

19 a 6:45 a.m., 6:45 p.m., 1850 b 2145, 10 p.m., 2230
c 0910, 9:20 a.m., 9:35 a.m. d 1300, 1:35 p.m., 1340
e 1640, 4:55 p.m., 1705 f 0800, 8:15 a.m., 2022
g 1400, 1428, 2:30 p.m. h 0640, 6 p.m., 1846
i 0830, 8 p.m., 2005 j 10 a.m., 10 p.m., 1015
k 0700, 7:25 a.m., 1920 l 3:15 p.m., 1530, 1535
m 8:15 a.m., 8 p.m., 2015 n 0400, 4:50 a.m., 1600

Strand: Chance and Data

Page 69

1 a 31 b 27 c 56 d 31 e 36 f 56
g 47 h 71 i 54 j 78 k 30 l 86
m 42 n 30 o 82

2 a 21 b 31 c 17 d 84 e 53 f 23
g 83 h 65 i 96 j 141 k 208 l 74
m 34 n 57

3 a 4 b 6 c 11 d 8 e 32 f 13
g 22 h 51

4 a 46 b 32 c 76 d 59 e 27 f 34

Page 70

1 a 2769 b 780 c 1188 d 425 e 1440 f 1152
g 1517 h 600 i 1870 j 1248

2 a 96 b 59 c 123 d 289 e 15 f 41
g 42 h 104 i 49 j 206

3 a 17.7 b 44.4 c 27.6 d 29 e 27.6 f 27.3
g 57.6 h 74.7 i 33.5 j 35.2 k 38.5 l 83.7
m 52.8 n 48.6 o 19 p 144.2 q 47.4 r 178.5
s 184.8 t 98.5 u 155.2 v 51.6

4 a 56.9 b 53.35 c 37.89 d 63.88 e 201.12 f 93.52
g 120.3 h 202.24 i 479.45 j 152.25 k 304.55 l 99.32
m 117.96 n 148.11 o 493.8 p 211.61 q 143.22 r 143.8

Page 71

1 a 2800 b 7300 c 5600 d 8900 e 4300 f 1700
g 3100 h 7000 i 9200 j 7900 k 16 800 l 67 800
m 35 100 n 43 500 o 26 700 p 83 100 q 58 700 r 10 200
s 92 300 t 76 500 u 576 400 v 813 900 w 498 400 x 717 300
y 206 900

2 a 3000 b 7000 c 4000 d 8000 e 5000 f 59 000
g 33 000 h 19 000 i 73 000 j 42 000 k 11 000 l 91 000
m 52 000 n 27 000 o 84 000 p 218 000 q 408 000 r 678 000
s 932 000 t 125 000 u 377 000 v 812 000 w 557 000 x 116 000
y 935 000

3 a 415 000, 420 000 b 631 000, 630 000 c 296 000, 300 000
d 155 000, 150 000 e 5 840 000, 5 840 000 f 3 267 000, 3 270 000
g 1 842 000, 1 840 000 h 7 256 000, 7 260 000 i 9 112 000, 9 110 000
j 8 371 000, 8 370 000 k 2 455 000, 2 460 000 l 6 088 000, 6 090 000
m 5 164 000, 5 160 000 n 4 159 000, 4 160 000 o 8 371 000, 8 370 000
p 1 403 000, 1 400 000

4 a Any number from 675 000 to 675 499
b Any number from 466 500 to 467 499
c Any number from 215 500 to 216 499
d Any number from 786 500 to 787 499
e Any number from 322 500 to 323 499
f Any number from 911 500 to 912 499
g Any number from 154 500 to 154 999
h Any number from 846 500 to 847 499
i Any number from 258 500 to 259 499
j Any number from 137 500 to 138 499
k Any number from 595 000 to 595 499
l Any number from 413 500 to 414 499
m Any number from 385 500 to 386 499
n Any number from 622 500 to 623 499
o Any number from 705 500 to 706 499
p Any number from 270 500 to 271 499

Page 72

1 a 24 b 54 c 18 d 61 e 8 f 42
g 13 h 36 i 65 j 55 k 27 l 78
m 37 n 93 o 86 p 40 q 12 r 99

2 a 13.8 b 27.4 c 92.2 d 45.8 e 34.1 f 6.2
g 56.1 h 82.6 i 79.2 j 29.2 k 63.7 l 50.5
m 15.4 n 93.8 o 38.7 p 81.1 q 32.1 r 64.2

3 a 56.51 b 12.85 c 47.18 d 63.02 e 70.17 f 82.94
g 15.44 h 23.82 i 95.07 j 24.64 k 7.9 l 38.2
m 10.35 n 82.85 o 26.01 p 49.81 q 53.1 r 76.22

4 23.25, 22.816, 22.609, 23.395, 22.7, 23.056

5 17.67, 18.39, 18.47, 18.135, 18.2, 17.51

6 35.63, 35.58, 35.62, 35.61, 35.617, 35.635, 35.59, 35.578

7 a 8.8, 8.75 b 5.1, 5.09 c 9.6, 9.62 d 8.1, 8.08
e 6.3, 6.34 f 8, 7.95 g 4.1, 4.13 h 9.6, 9.58
i 5.2, 5.21 j 3.5, 3.46

Page 73

1 a K10.60 b K15.30 c K6.70 d K11.60 e K16.80 f K8.10
g K12.90 h K9.90 i K8.70 j K5.50 k K19.30 l K18.00
m K9.20 n K21.50 o K25.60 p K14.90 q K5.00 r K7.60
s K20.00 t K16.30 u K13.10 v K22.20 w K15.80 x K23.10

2 a K23.00 b K67.00 c K75.00 d K41.00 e K30.00 f K22.00
g K18.00 h K35.00 i K63.00 j K89.00 k K56.00 l K9.00
m K49.00 n K51.00 o K73.00 p K14.00 q K29.00 r K96.00
s K82.00 t K26.00 u K8.00 v K12.00 w K85.00 x K57.00

3 a 51t b 24t c 71t d 15t e 35t f 79t
g 46t h 67t i 37t j 53t k 10t l 91t
m 32t n 49t o 89t p 57t q 86t r 73t
s 19t t 32t u 60t v 43t w 27t x 43t

4 K24.50, K25.15, K24.85, K24.65, K25.37

5 K68.49, K68.45, K68.52, K68.51, K68.47

6 K112.87, K112.85, K112.92, K112.88

7 a K60.00 b K250.00 c K130.00 d K180.00 e K100.00 f K710.00
g K550.00 h K70.00 i K390.00 j K110.00 k K410.00 l K300.00
m K360.00 n K810.00 o K30.00 p K180.00 q K960.00 r K900.00
s K520.00 t K280.00 u K650.00 v K480.00 w K70.00 x K340.00
y K500.00

Page 74

1 a 26° b 24° c 28° d 22° e 31° f 29°
g 34° h 22° i 38° j 35° k 17° l 21°
m 40° n 32° o 20° p 23° q 26° r 39°

2 a 15 s b 12 s c 25 s d 24 s e 13 s f 21 s
g 9 s h 23 s i 13 s j 7 s k 16 s l 30 s
m 12 s n 10 s o 22 s p 18 s q 24 s r 8 s

3 a 11.8 s b 15.1 s c 12.6 s d 9.5 s e 15.9 s f 13.4 s
g 17.6 s h 16.1 s i 14.2 s j 8 s k 10.7 s l 11.8 s
m 16.9 s n 12.7 s o 18.1 s p 23.5 s q 21.8 s r 19.4 s
s 14.7 s t 10.2 s u 20.5 s v 24.2 s w 17.9 s x 16 s

4 a 9 L b 18 L c 10 L d 5 L e 4 L f 10 L
g 15 L h 23 L i 3 L j 8 L k 12 L l 6 L
m 9 L n 1 L o 5 L p 1 L q 8 L r 3 L
s 1 L t 3 L u 1 L v 5 L w 3 L x 7 L

5 a 5.4 L b 2.2 L c 3.1 L d 0.8 L e 4.5 L f 2.9 L
g 1.2 L h 7.9 L i 5.1 L j 1.8 L k 3.5 L l 0.4 L
m 2.7 L n 8.1 L o 1.6 L p 4 L q 5.8 L r 3.1 L
s 0.8 L t 2.7 L u 6.3 L v 9.1 L w 1 L x 4.9 L

6 a 6 kg b 4 kg c 2 kg d 1 kg e 6 kg f 1 kg
g 8 kg h 6 kg i 4 kg j 5 kg k 3 kg l 5 kg
m 2 kg n 2 kg o 4 kg p 9 kg q 3 kg r 1 kg
s 7 kg t 2 kg u 2 kg v 10 kg w 7 kg x 4 kg

7 a 2.5 t b 4.8 t c 0.2 t d 1.7 t e 5.7 t f 2.1 t
g 1.5 t h 3.8 t i 8 t j 9.2 t k 4.1 t l 0.6 t
m 5.6 t n 1.1 t o 2.4 t p 0.7 t q 3.5 t r 6.7 t

Page 75 Assessment

1 C 2 A 3 A 4 D 5 B
6 C 7 D 8 B 9 A 10 D
11 A 12 C 13 B 14 A 15 D
16 B 17 C 18 B 19 D

Strand: Patterns and Algebra

Page 76

1 a 26, 25, 29 b 20, 15, 9 c 17, 22, 28, 35
d 60, 70, 50 e 32, 64, 128, 256 f 63, 127, 255, 511
g 33, 32, 27 h 21, 23, 28 i 12, 9, 8
j 18, 19, 23 k 36, 35, 45 l 19, 21, 24
m 16, 18, 21 n 33, 66, 65 o 48, 96, 192, 384
p 54, 108, 110 q 243, 729, 2187 r 67, 57, 56
s 28, 36, 45 t 36, 49, 64, 81

2 a Add 6 to each number b Subtract 7 from each number
c Subtract 10, subtract 5 … d Add 1, add 2, add 3, add 4 …
e Double each number f Multiply by 5
g Halve each number h Add 5, subtract 1 …
i Double, subtract 1 … j Subtract 1, subtract 2, subtract 3 …
k Add two consecutive numbers to find the next number in the sequence
l Add 8, subtract 1 … m Multiply by 5, add 1 …
n Subtract 9, subtract 8, subtract 7 … o Add 9, subtract 1 …
p Double, add 1 … q Subtract 10, subtract 3 …
r Subtract 2, subtract 5 … s Add 1.5, add 2 …
t Subtract 1.25, subtract 1.5 …

3 a–r Teacher to check

Page 77

1

IN	OUT
11	15
16	20
27	31
33	37
59	63
88	92

2

IN	OUT
22	15
26	19
35	28
40	33
58	51
83	76

3

IN	OUT
4	12
9	27
12	36
20	60
25	75
30	90

4

IN	OUT
25	5
40	8
50	10
100	20
250	50
325	65

5

IN	OUT
6	48
9	72
11	88
15	120
21	168
40	320

6

IN	OUT
12	27
23	38
36	51
47	62
69	84
86	101

7

IN	OUT
18	3
30	5
54	9
66	11
90	15
108	18

8

IN	OUT
21	9
40	28
65	53
73	61
96	84
127	115

9

IN	OUT
5	13
8	19
12	27
15	33
30	63
45	93

10

IN	OUT
10	15
25	60
28	69
30	75
45	120
67	186

11

IN	OUT
16	9
28	15
42	22
70	36
92	47
114	58

12

IN	OUT
7	9
13	11
22	14
31	17
37	19
46	22

13

IN	OUT
11	80
17	140
32	290
47	440
61	580
85	820

14

IN	OUT
20	5
46	18
58	24
72	31
106	48
300	145

15

IN	OUT
6	44
11	74
15	98
23	146
32	200
40	248

Page 78

1 Double: 28, 60, 50

2 Double and add 3: 21, 33, 23

3 Double and subtract 1: 15, 9, 21

4 Multiply by 4 and add 1: 33, 9, 49

5 Halve and add 1: 7, 6, 13

6 Halve and add 2: 8, 27, 11

7 Multiply by 3 and subtract 1:59, 32, 26

8 Multiply by 10 and subtract 10: 10, 160, 220

9 Multiply by 5 and add 10: 50, 65, 110

10 Double and add 2: 202, 44, 34

11 Divide by 4 and add 1: 13, 26, 2

12 Multiply by 5 and add 1: 56, 101, 126

13 Multiply by 3 and add 10: 43, 55, 100

14 Multiply by itself and add 1: 50, 26, 145

15 Multiply by 3 and subtract 5: 55, 67, 115

Pages 79–80

1 **a** $x = y + 5$ **b** $m = n - 10$ **c** $a = 3b$ **d** $f = g \div 5$
e $s = t + 8$ **f** $k = p - 4$ **g** $w = 5x$ **h** $c = 12 \div d$
i $b = g - 2$ **j** $t = f + 4 + 3$ **k** $m = 4n$ **l** $y = x \div 7$

2 **a** *d* is the same as 6 added to *x*
b *c* is the same as *g* minus 8
c *m* is the same as *n* multiplied by 11
d *k* is the same as *s* multiplied by 3
e *y* is the same as 20 divided by *x*
f *a* is the same as *b* plus 6 and 7
g *x* is the same as *y* minus 12
h *f* is the same as *g* divided by 2
i *p* is the same as 5 plus *q*
j *s* is the same as 5 times *t*
k *b* is the same as 3 plus *a* plus 7
l *w* is the same as *z* minus 8 minus 3

3 **a** $m = 5(n + 7)$ **b** $p = 10(q - 8)$ **c** $a = 9b + 4$ **d** $d = c \div 2 - 6$
e $w = 7z - 3$ **f** $g = 12(h + 6)$ **g** $x = y \div 4 + 10$ **h** $k = (j - 6) \div 4$
i $s = 3(t + 15)$ **j** $f = g \div 8 - 11$ **k** $b = 6a + 7$ **l** $y = 5(x - 12)$

4 **a** To get *g*, first multiply *h* by 4 and then subtract 6.
b To get *x*, first add 4 to *y* and then multiply by 5.
c To get *m*, first divide *n* by 2 and then add 3.
d To get *a*, first subtract 7 from *b* and then multiply by 6.
e To get *d*, first subtract 4 from *c* and then divide by 3.
f To get *s*, first multiply *t* by 2 and then add 9.
g To get *k*, first add 8 to *j* and then multiply by 11.
h To get *f*, first add *g* to 8 and then subtract 5.
i To get *w*, first multiply *z* by 7 and then add 10.
j To get *h*, first subtract 8 from *k* and then multiply by 3.
k To get *p*, first add *q* to 5 and then divide by 2.
l To get *y*, first multiply *x* by 8 and then add 9.

Pages 80–81

1 **a** $b = 3$ **b** $y = 4$ **c** $p = 5$ **d** $m = 7$ **e** $x = 12$ **f** $p = 15$
g $a = 14$ **h** $d = 27$ **i** $c = 11$ **j** $g = 63$ **k** $f = 24$ **l** $n = 64$
m $s = 39$ **n** $q = 27$ **o** $k = 15$ **p** $t = 25$ **q** $f = 16$ **r** $z = 36$
s $h = 83$ **t** $p = 9$ **u** $r = 25$ **v** $n = 7$ **w** $e = 7$ **x** $g = 34$

2 **a** $y = 12$ **b** $a = 8$ **c** $n = 10$ **d** $g = 11$ **e** $x = 31$ **f** $b = 28$
g $m = 30$ **h** $k = 19$ **i** $w = 20$ **j** $p = 9$ **k** $t = 5$ **l** $g = 14$
m $d = 9$ **n** $z = 30$ **o** $y = 10$ **p** $f = 6$ **q** $a = 5$ **r** $s = 48$
s $k = 9$ **t** $d = 3$ **u** $p = 7$ **v** $h = 3$ **w** $r = 20$ **x** $m = 3$

3 **a** $x = 5$ **b** $y = 2$ **c** $s = 3$ **d** $p = 21$ **e** $z = 5$ **f** $a = 2$
g $t = 25$ **h** $b = 7$ **i** $d = 6$ **j** $w = 24$ **k** $g = 20$ **l** $f = 60$
m $h = 12$ **n** $x = 40$ **o** $y = 10$ **p** $n = 20$ **q** $m = 18$ **r** $v = 5$
s $k = 12$ **t** $b = 9$ **u** $s = 20$ **v** $p = 18$ **w** $r = 10$ **x** $u = 17$

4 **a** $x = 1$ **b** $n = 10$ **c** $a = 10$ **d** $y = 2$ **e** $g = 10$ **f** $b = 5$
g $m = 3$ **h** $p = 4$ **i** $p = 5$ **j** $d = 16$ **k** $h = 2$ **l** $s = 12$
m $w = 8$ **n** $c = 11$ **o** $d = 18$ **p** $z = 11$ **q** $f = 3$ **r** $k = 3$
s $x = 2$ **t** $y = 11$ **u** $g = 30$ **v** $a = 13$ **w** $m = 50$ **x** $q = 7$

Pages 82–83 Assessment

1 **a** 34, 31, 30 **b** 18, 17, 21 **c** 23, 46, 47
d 48, 46, 45 **e** 42, 44.5, 46 **f** 19, 18.5, 16.5

2 **a** Subtract 6 from each number **b** Add 1, add 5 …
c Add 2, subtract 5 … **d** Double, add 2.5 …
e Subtract 0.5, subtract 1 … **f** Multiply by 3, add 1 …

3 **a**

IN	OUT
3	12
8	32
10	40
12	48
20	80
50	200

b

IN	OUT
6	2
12	4
30	10
33	11
90	30
150	50

c

IN	OUT
8	21
15	35
20	45
24	53
32	69
40	85

d

IN	OUT
10	2
18	6
22	8
30	12
50	22
60	27

4 **a** Multiply by 5 **b** Halve and subtract 1
c Double and add 2 **d** Square (multiply by itself)

5 **a** $x = y + 10$ **b** $f = 4g$ **c** $m = n - 6$
d $a = b \div 12$ **e** $s = 5t$ **f** $w = z - 8$

6 **a** $x = 6$ **b** $p = 9$ **c** $f = 36$ **d** $y = 15$ **e** $a = 7$ **f** $b = 2$
g $w = 5$ **h** $d = 6$ **i** $g = 4$ **j** $k = 5$ **k** $h = 4$ **l** $c = 5$